Advanced Topics in Physics for Undergraduates

Advanced Topics in Physics for Undergraduates explores classical mechanics, electrodynamics, and quantum mechanics beyond the standard introductory courses. Designed to support departments with limited resources, this book integrates these advanced topics into a single, cohesive volume, offering students a unified perspective on fundamental physical principles. By presenting these interconnected subjects in one voice, it provides a compact yet comprehensive resource that enhances understanding and bridges the gaps between core physics disciplines.

Features:

- A structured three-part approach covering classical mechanics, electrodynamics, and quantum mechanics
- In-depth exploration of Lagrange and Hamilton formalisms, small oscillations, conservation principles, scalar and vector potentials, radiation, and special relativity
- Advanced quantum mechanics topics such as perturbation theory, scattering, quantum information, and quantum computing

This book serves as an invaluable guide for undergraduate students seeking to deepen their knowledge of physics, preparing them for further academic study or careers in physics and related fields. Its clear explanations and structured approach make it accessible to learners looking to advance their understanding beyond traditional coursework.

Asim Gangopadhyaya earned his Ph.D. in theoretical particle physics from the City University of New York. He is an accomplished educator and a recipient of the Sujack Award for Teaching Excellence at Loyola University Chicago. His teaching extends beyond the classroom, as he frequently collaborates with undergraduate students on both pedagogical and research articles, preparing them to actively contribute to academic scholarship. Prof. Gangopadhyaya is also a co-author of a book on supersymmetric quantum mechanics. His current research focuses on the solvability of quantum mechanical systems, an active area of research in theoretical physics.

Constantin Rasinariu holds a Ph.D. in theoretical physics from the University of Illinois at Chicago, where his research focused on superintegrable systems. A dedicated educator and researcher, he brings over three decades of teaching experience across two continents. Prof. Rasinariu has co-authored two textbooks: one on mathematical methods and another one, with Prof. Gangopadhyaya, on supersymmetric quantum mechanics, now in its second edition. His current research interests include supersymmetric quantum mechanics, the fundamentals of quantum mechanics, and quantum computing.

Advanced Topics in Physics for Undergraduates

Asim Gangopadhyaya and Constantin Rasinariu

CRC Press
Taylor & Francis Group
Boca Raton London New York

CRC Press is an imprint of the
Taylor & Francis Group, an **informa** business

Designed cover image: Asim Gangopadhyaya and Constantin Rasinariu

First edition published 2026
by CRC Press
2385 NW Executive Center Drive, Suite 320, Boca Raton FL 33431

and by CRC Press
4 Park Square, Milton Park, Abingdon, Oxon, OX14 4RN

CRC Press is an imprint of Taylor & Francis Group, LLC

ISBN: 978-0-367-77582-7 (hbk)
ISBN: 978-1-032-00632-1 (pbk)
ISBN: 978-1-003-17496-7 (ebk)

DOI: 10.1201/ 9781003174967

Typeset in CMR10 Modern
by KnowledgeWorks Global Ltd.

Publisher's note: This book has been prepared from camera-ready copy provided by the authors.

Access the Support Material: www.routledge.com/9780367775827

AG: To Alpana

CR: To Mariana

Contents

Preface

This book is an outcome of our extensive experience in teaching undergraduate students at Loyola University Chicago. It emerged out of our need to provide students an introduction to material generally covered in second tier courses in areas such as Classical Mechanics, Electricity and Magnetism and Quantum Mechanics, when constraints prevented us from offering a comprehensive course on each subject.

Many physics departments encounter similar constraints, rendering it impractical to offer "second courses" in these areas due to limited faculty, smaller class sizes, or other constraints. At Loyola, we consistently teach the mandatory first course of the sequence in each of these core subjects. However, the absence of required advanced courses has left some students without the opportunity to broaden their knowledge.

This book is designed to address this gap by providing a comprehensive capstone resource for upper-level undergraduates. Instead of using separate textbooks for three distinct courses, our book proposes a single course that integrates advanced topics in mechanics, electricity and magnetism, and quantum mechanics into a comprehensive volume. The aim is to provide a fifteen-week course that covers material typically not included in the first courses of the sequence, utilizing a consistent set of mathematical tools and notation.

In developing this resource, we have drawn from our experiences in team-taught courses that covered advanced topics across these three domains. We hope the present textbook addresses a common challenge: the fragmented learning experience resulting from multiple textbooks with varying approaches and notations. By presenting these interconnected subjects in one voice, we hope to provide students with a more cohesive experience, helping them build a deeper understanding of the physical principles that unify these fields.

The book is divided into three parts, covering topics in mechanics, electricity and magnetism, and quantum mechanics.

Part I assumes that students are familiar with introductory mechanics such as solving dynamical problems using Newton's laws of motion, and the Lagrangian and Hamiltonian formalisms. This is usually the case after completing an introductory course in theoretical, analytical, or classical mechanics. We conclude this part with an introduction to small oscillations, an area of significant importance in physics.

Due to various teaching approaches of electricity & magnetism, and quantum mechanics, parts II and III begin with brief reviews of each field to unify notation.

In part II, we assume that students took a junior-level course in electricity & magnetism and are familiar with Coulomb's law, Gauss' law, and Ampère's law. They should also have experience with Laplace's equation and its solutions in various coordinate systems. Therefore, in this part, we introduce more advanced topics such as electrodynamics, conservation principles, scalar and vector potentials, radiation, and special relativity. If time is limited, instructors could skip the chapter on radiation without affecting the readability of later chapters.

Finally, the last part deals with topics in quantum mechanics. Here, we assume that students are familiar with the postulates of quantum mechanics, Dirac's bra-ket notation, as well as solving systems such as the infinite and finite potential well, the harmonic oscillator, and Coulomb potential. In part III we chose to introduce elements of time-independent and time-dependent perturbation theory, scattering theory, and quantum computing. The ideas of quantum information, quantum encryption, and quantum computation are likely to become central in physics and technology in the coming years. If time is limited, instructors could skip the chapter on scattering.

This book is designed to support departments with limited resources, giving students access to a curriculum that would otherwise require multiple courses. We hope it will serve as a comprehensive and approachable guide for students looking to expand their understanding of advanced physics topics, fostering a level of preparation that will benefit them as they move into further studies or careers in physics and related fields.

Chicago, November 2024

Part I

Classical Mechanics

1

Lagrange and Hamilton Formalism

1.1 Introduction

Lagrange formalism together with the related Hamilton formalism are incredibly powerful and far-reaching tools for studying complex physical systems. In classical mechanics, Lagrange and Hamilton equations are equivalent to Newton's laws, and they are quite useful especially when solving for cases where forces acting on a system are difficult to identify. Although developed as equivalent ways of solving problems in classical mechanics, Lagrange and Hamilton equations became essential in other areas, such as electromagnetism, quantum mechanics, and quantum field theory.

While we hope that the reader is already familiar with Lagrange and Hamilton equations, it is instructive to present here a short introduction. We start with the concept of generalized coordinates and the number of degrees of freedom needed to describe a system.

1.2 Generalized coordinates and constraints

Generalized coordinates constitute a set of variables which completely describes the configuration of a system. They change as the system evolves under the action of forces. While the particular choice of these coordinates (usually chosen as positions and angles) is entirely up to us, the minimum number of such variables needed to completely describe the system is fixed and depends only on the system. For example, the motion of a point-like mass is fully described by three coordinates $(x(t), y(t), z(t))$ in Cartesian coordinates, or $(r(t), \theta(t), \phi(t))$ in spherical coordinates. In this case we need three coordinates to describe the system. When unspecified, we denote these generalized variables as q_i and often omit their time dependence. A system of N unconstrained particles is described by $3N$ generalized coordinates. But, if we introduce a constraint in the system, say two of the particles are connected by a rigid bar, then we only need $3N-1$ coordinates to describe its motion. For k constraints, it is enough to know the functional form of $n = 3N - k$ generalized coordinates, in order to get the full picture of system's evolution. The

DOI: 10.1201/9781003174967-1

minimum number of generalized coordinates needed to completely describe a system is known as the *number of degrees of freedom.*

Example 1.1 The degrees of freedom of a rigid body
A brick is a good example of a rigid body. Consider a brick constrained to move on a flat table. It has three degrees of freedom: two translations (q_1 and q_2) and one rotation (q_3). Now, let us consider the same brick, but this time we allow it to move freely in space. It has six degrees of freedom: three describing the motion of its center of mass (q_1, q_2, and q_3) and three rotations (q_4, q_5, and q_6) as illustrated in Fig. 1.1.

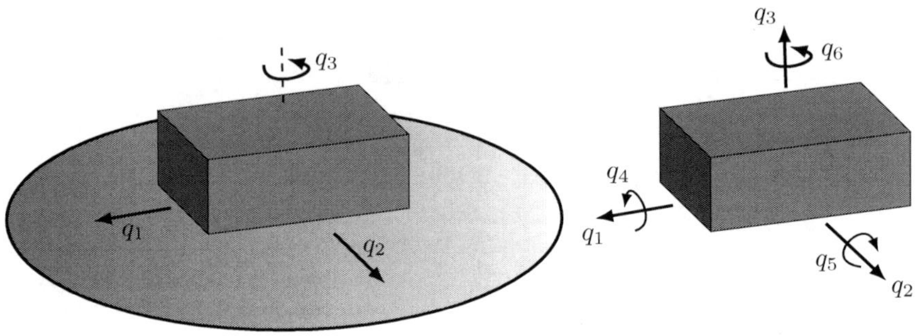

FIGURE 1.1
Systems with different degrees of freedom: in the first case the brick is constrained to move on the two-dimensional plane, while in the second case it is free to move in the three-dimensional space.

Let us consider a system of N particles with n degrees of freedom. If each coordinate can vary independently of the other, the system is said to be *holonomic.* Holonomic constraints can be expressed as algebraic relations among the coordinates

$$f_i(x_1, \ldots, x_{3N}; t) = 0 \; ; \quad 1 = 1, 2, \ldots, k \; . \tag{1.1}$$

In such cases, we can always find a set of n independent generalized coordinates $q_1, q_2, \ldots, q_n$ that completely describe the configuration of the system:

$$q_j = q_j(x_1, \ldots, x_{3N}; t) \; , \quad j = 1, 2, \ldots, n \; . \tag{1.2}$$

If the constraints (1.1) depend explicitly on time, they are called *rheonomic,* otherwise *scleronomic.*

There are cases in which the particles' velocities appear in the expressions for the constraints

$$f_i(x_1, \ldots, x_{3N}; \dot{x}_1, \ldots, \dot{x}_{3N}; t) = 0 \; , \quad i = 1, 2, \ldots, k \; . \tag{1.3}$$

If these differential equations cannot be integrated, then the constraints are nonholonomic.

Determining the correct number of degrees of freedom is crucial in solving practical problems. In the next section, we will introduce the Lagrange equations and explicitly show how it is used to describe the motion of a system.

1.3 Lagrange equations

For simplicity, let us consider the case of a particle in a one-dimensional motion. Its motion can be described by $x(t)$ or by a generalized coordinate $q(t)$, which is related to $x(t)$ via

$$q = q(x(t), t) \; . \tag{1.4}$$

The inverse transformation is

$$x = x(q(t), t) \; . \tag{1.5}$$

Then, the velocity $\dot{x} = dx/dt$ becomes

$$\dot{x} = \frac{\partial x}{\partial q}\dot{q} + \frac{\partial x}{\partial t} \tag{1.6}$$

Note that because x doesn't depend explicitly on $\dot{q}$, from Eq. (1.6), we have

$$\frac{\partial \dot{x}}{\partial \dot{q}} = \frac{\partial x}{\partial q} \; , \tag{1.7}$$

which will be useful in a moment.

Newton's equation of motion

$$\dot{p} = F(x, \dot{x}, t), \tag{1.8}$$

contains the force $F(x, \dot{x}, t)$, which in many cases could be complicated or difficult to identify. Lagrange formalism bypasses this difficulty by using the kinetic and potential energies of the system, which being scalar are often easier to find.

We start by writing the momentum $p = m\dot{x}$ of a particle in terms of its kinetic energy $T = \frac{1}{2}m\dot{x}^2$:

$$p = \frac{dT}{d\dot{x}} \; . \tag{1.9}$$

Next, in analogy with Eq. (1.9), we define the *generalized momentum* as

$$\tilde{p} \equiv \frac{\partial T(q, \dot{q}, t)}{\partial \dot{q}} \; , \tag{1.10}$$

where $\dot{q}$ is the *generalized velocity*. This generalized momentum is also known as the *canonical momentum* conjugate to q. The kinetic energy $T(q, \dot{q}, t)$ is $T(\dot{x})$ with $\dot{x}$ given by Eq. (1.6). Later on, we will drop the $\tilde{p}$ notation and write simply p for the generalized momentum. Since x does not depend on $\dot{q}$, using the chain differentiation, we obtain

$$\tilde{p} = \frac{dT}{d\dot{x}}\frac{\partial \dot{x}}{\partial \dot{q}} = \frac{dT}{d\dot{x}}\frac{\partial x}{\partial q} , \tag{1.11}$$

where we used Eq. (1.7) in the last step. Therefore

$$\tilde{p} = p\frac{\partial x}{\partial q} , \tag{1.12}$$

which we now differentiate with respect to time

$$\dot{\tilde{p}} = \dot{p}\frac{\partial x}{\partial q} + p\frac{d}{dt}\frac{\partial x}{\partial q} = F\frac{\partial x}{\partial q} + p\frac{\partial \dot{x}}{\partial q} . \tag{1.13}$$

To justify the differential interchange of $\frac{d}{dt}$ and $\frac{\partial}{\partial q}$ in the second step above, we can write

$$\frac{d}{dt}\frac{\partial x}{\partial q} = \frac{\partial}{\partial q}\frac{\partial x}{\partial q}\dot{q} + \frac{\partial}{\partial t}\frac{\partial x}{\partial q} = \frac{\partial}{\partial q}\frac{\partial x}{\partial q}\dot{q} + \frac{\partial}{\partial q}\frac{\partial x}{\partial t} = \frac{\partial \dot{x}}{\partial q} . \tag{1.14}$$

Here we must note that in the Lagrangian formalism, q and $\dot{q}$ are regarded as independent variables, i.e., $\partial\dot{q}/\partial q = 0$.

The term $Q = F\partial x/\partial q$ is called the *generalized force*. The force F can be separated into a part that can be expressed in terms of the potential energy $V(x)$ as $-dV(x)/dx$ and a part F_1 that cannot be expressed in terms of a potential energy. Then, Eq. (1.13) becomes

$$\dot{\tilde{p}} = -\frac{dV(x)}{dx}\frac{\partial x}{\partial q} + F_1\frac{\partial x}{\partial q} + \frac{\partial T}{\partial \dot{x}}\frac{\partial \dot{x}}{\partial q} = \frac{\partial T}{\partial q} - \frac{\partial V(q)}{\partial q} + Q_1 \tag{1.15}$$

where $Q_1 = F_1\frac{\partial x}{\partial q}$. The function $V(q)$ has the same value at each point as $V(x(q))$. It is a different function than $V(x)$, but we keep the same notation for simplicity. Therefore, Eq. (1.15) yields

$$\dot{\tilde{p}} = \frac{\partial}{\partial q}(T(\dot{q}, q, t) - V(q)) + Q_1 . \tag{1.16}$$

Because $\partial V/\partial\dot{q} = 0$, we can write Eq. (1.10) as

$$\tilde{p} = \frac{\partial}{\partial \dot{q}}\left(T(q, \dot{q}, t) - V(q)\right) . \tag{1.17}$$

Now, taking the time derivative of (1.17) and comparing it with (1.16), we finally obtain

$$\frac{d}{dt}\left(\frac{\partial L}{\partial \dot{q}}\right) - \frac{\partial L}{\partial q} = Q_1 , \tag{1.18}$$

where we introduced the *Lagrangian* function of the system L:

$$L(\dot{q}, q, t) \equiv T(\dot{q}, q, t) - V(q) \; . \tag{1.19}$$

If all forces F_1 can be derived from the potential energy, then $Q_1 = 0$ and we obtain the Lagrange equation

$$\frac{d}{dt}\left(\frac{\partial L}{\partial \dot{q}}\right) - \frac{\partial L}{\partial q} = 0 \; . \tag{1.20}$$

In general, let us consider a system with n degrees of freedom, described by the generalized coordinates q_i, $i = 1, 2, \ldots, n$. As in the one-dimensional case, the generalized velocities are the time derivatives of q_i and are denoted by $\dot{q}_i$.

Because $q_i(0)$ and $\dot{q}_i(0)$ can be independently chosen, it follows that $q_i(t)$ and $\dot{q}_i(t)$ are independent variables, in the Lagrangian formalism. Then, the Lagrangian of the system is given by

$$L(\dot{q}_i, q, t) = T(\dot{q}_i, q_i, t) - V(q_i) \; . \tag{1.21}$$

and the Lagrange equations of motion are

$$\frac{d}{dt}\left(\frac{\partial L}{\partial \dot{q}_i}\right) - \frac{\partial L}{\partial q_i} = 0 \; , \quad i = 1, \ldots, n \; . \tag{1.22}$$

Here n is the number of the degrees of freedom of the system. Equations (1.22) are also known as *Euler-Lagrange equations*, and they can be derived from a variational principle, the principle of the least action, see for example [16].

Thus, regardless of how complicated the system appears, all we have to do is to find a single scalar function, the Lagrangian of the system $L(q_i, \dot{q}_i, t)$. Then, the equations of motion follow immediately from Eq. (1.22). Solving these equations gives us the complete mechanical description of the evolution of the system.

The following examples illustrate the Lagrangian formalism starting from a very simple case.

Example 1.2 The free-falling body
Find the motion of a free-falling point-like particle of mass m using the Lagrangian formalism.

In this case, we can completely describe the motion of the system using one coordinate $q = z$, which is the height above the ground. The kinetic energy is $T = m\dot{z}^2/2$ and the potential energy is $V = mgz$, where we considered the ground to be at zero potential energy. Then, the Lagrangian of the system is

$$L(z, \dot{z}) = \frac{1}{2} m\dot{z}^2 - mgz \; . \tag{1.23}$$

Because $N = 1$, there is only one Lagrange equation, which after simplification reads

$$\ddot{z} - g = 0 \tag{1.24}$$

Its solution, $z(t) = z(0) + \dot{z}(0)\,t + \frac{1}{2}gt^2$, describes the well-known free-fall motion.

Example 1.3 The double pendulum
Let us consider the case of the ideal double pendulum pictured in Fig. 1.2, which can oscillate only in the vertical plane. Find the equations of motion for this system.

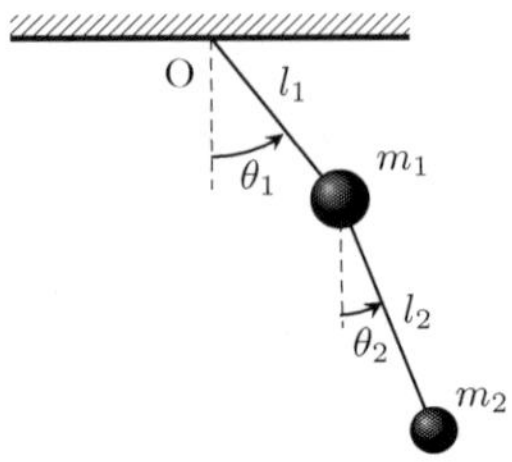

FIGURE 1.2
Double pendulum: the suspension point of the second pendulum is connected to the bob of the first pendulum.

In order to write the Lagrange equations, we first determine the number of degrees of freedom n for this system. A system of two particles, free to move in a plane, has four degrees of freedom (two for each particle). In this case the two ideal strings constrain further the motion of the two masses. Hence, we obtain that this system has $n = 2$ degrees of freedom. We will see that we can choose θ_1 and θ_2 as generalized coordinates. It is easy to express the positions of the two masses first in Cartesian coordinates (x, y), where x-axis is chosen horizontally with the origin in O, and the y-axis is chosen along the vertical direction, pointing down. We obtain for the Lagrangian $L = T - V$

$$L = \frac{m_1}{2}(\dot{x_1}^2 + \dot{y_1}^2) + \frac{m_2}{2}(\dot{x_2}^2 + \dot{y_2}^2) + m_1 g y_1 + m_2 g(y_1 + y_2) \,, \tag{1.25}$$

where we have considered the potential energy to be zero at point O. Then, from the geometry of the problem we have $x_1 = l_1 \sin\theta_1$, $y_1 = l_1\cos\theta_1$, $x_2 = l_1\sin\theta_1 + l_2\sin\theta_2$, and $y_2 = l_1\cos\theta_1 + l_2\cos\theta_2$. Consequently, the Lagrangian becomes

$$\begin{aligned} L = & \frac{(m_1 + m_2)\, l_1^2}{2}\dot{\theta_1}^2 + \frac{m_2\, l_2^2}{2}\dot{\theta_2}^2 + \dot{\theta_1}\dot{\theta_2}\, m_2\, l_1 l_2 \cos(\theta_1 - \theta_2) \\ + & (m_1 + m_2)\, g l_1 \cos\theta_1 + m_2\, g\, l_2 \cos\theta_2 \,. \end{aligned} \tag{1.26}$$

Because the system is described by two generalized coordinates θ_1 and θ_2, we have two Euler-Lagrange equations

$$\frac{d}{dt}\left(\frac{\partial L}{\partial \dot{\theta}_1}\right) - \frac{\partial L}{\partial \theta_1} = 0 \tag{1.27}$$

$$\frac{d}{dt}\left(\frac{\partial L}{\partial \dot{\theta}_2}\right) - \frac{\partial L}{\partial \theta_2} = 0 \ , \tag{1.28}$$

which after simplifications give

$$\begin{aligned} &\ddot{\theta}_1 \left(1 + \frac{m_1}{m_2}\right) l_1 + \ddot{\theta}_2\, l_2 \cos(\theta_1 - \theta_2) + \dot{\theta}_2^{\,2}\, l_2 \sin(\theta_1 - \theta_2) \\ &\qquad + \left(1 + \frac{m_1}{m_2}\right) g \sin\theta_1 = 0 \ , \\ &\ddot{\theta}_2\, l_2 + \ddot{\theta}_1\, l_1 \cos(\theta_1 - \theta_2) - \dot{\theta}_1^{\,2}\, l_1 \sin(\theta_1 - \theta_2) + g \sin\theta_2 = 0 \ . \end{aligned} \tag{1.29}$$

In this form, the two coupled non-linear differential equations cannot be solved analytically. In the next chapter we will see that there is a beautiful way of studying these types of motion, by considering the limiting case of small angles. Also, note that the second equation doesn't depend on m_2. Can you think of a reason for this behavior?

Exercise 1.1 Derive equations (1.29).

The next example is a bit more involved. As expected, the Lagrangian approach will yield a straightforward answer, without the need to consider the forces between the different subsystems.

Example 1.4 Hanging a coat over the corner of a table

Anyone who has tried to hang a coat on a corner of a table knows the futility of such an endeavor. In fact, for a frictionless rectangular table the coat will slide toward the corner even if that corner is tilted upward. To tackle this problem, we will consider [41] a simple mechanical model that emulates the motion of the coat on the corner of a table. This example consists of a massless string that connects two hanging masses, placed symmetrically over a corner.

Concretely, let us consider two equal masses A and B are connected via an ideal string of length d which is symmetrically placed over the corner of a frictionless horizontal table, depicted in Fig. 1.3. The center point of the string is labeled P. Initially, the system is at rest, with the masses at the table height and horizontally separated by the length d of the string.

As the system is released and point P advances toward the corner, we are interested in finding the motion of P as a function of time and the angle α subtended by the edges of the table at the corner O (see Fig. 1.3). The motion of the two masses and of the center of mass is left as an exercise for the reader.

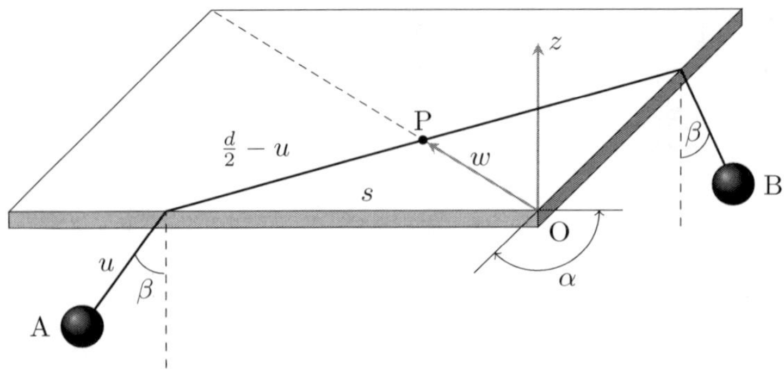

FIGURE 1.3
Two identical masses A and B connected by an ideal string of length d, falling symmetrically over the corner of a horizontal table. Note that the point P moves toward point O, opposite to the directed segment $\vec{OP}$.

Note an interesting aspect of this problem: the string moves toward the corner even if the table is tiled upwards by a given angle. Of course, in all cases, the center of mass of the system moves down.

The central point P moves along the bisector of angle α, which we take to be the w-axis with the origin at point O (Fig. 1.3). Mass A moves along the side s of the table and vertically along z, with mass B moving symmetrically along the other edge of the table. The center of mass of the system moves along the w-axis towards O and also moves down along the z-axis. Initially, the center of mass of the system is at the level of the table, at $z = 0$. Some time after release, as depicted in Fig. 1.3, masses A and B have descended by a vertical distance $u \cos\beta$ as point P moves along w towards the vertex O. Here β is the angle that the hanging masses make with the vertical, which is indicative of the accelerations of the masses. In Appendix 1.8 we show that this angle is constant and is given by

$$\beta = \frac{\alpha}{2} \; . \tag{1.30}$$

Now we are in position to write the equations of motion using the Lagrangian formalism. From the geometry of the system (see Fig. 1.3), we have

$$u = d/2 - w \tan\beta \tag{1.31}$$

and

$$s = w \sec\beta \; . \tag{1.32}$$

Mass A moves along the side s with speed $\dot{s}$ and along u with speed $\dot{u}$, so its velocity vector is $\vec{v}_A = \dot{s}\,\hat{\ell} + \dot{u}\,\hat{t}\,'$, where $\hat{\ell}$ and $\hat{t}\,'$ are the corresponding unit vectors (see also Fig. 1.12 in Appendix 1.8). Since A and B move symmetrically, they have the same kinetic energy, and thus the total kinetic energy of the system is given by

$$\begin{aligned} T &= 2\left(\frac{m}{2}\right)\left[\dot{s}^2 + \dot{u}^2 + 2|\dot{s}||\dot{u}|\cos\left(\pi/2 + \beta\right)\right] \\ &= 2\left(\frac{m}{2}\right)\left(\dot{s}^2 + \dot{u}^2 - 2|\dot{s}||\dot{u}|\sin\beta\right) \\ &= m\dot{w}^2\left(\sec^2\beta + \tan^2\beta - 2\tan^2\beta\right) \\ &= m\dot{w}^2 \ . \end{aligned} \tag{1.33}$$

Similarly, taking the surface of the table as the zero point, the potential energy of the system is

$$\begin{aligned} V &= -2mgu\cos\beta \\ &= -2mg\left(d/2 - w\tan\beta\right)\cos\beta \\ &= 2mgw\sin\beta - mgd\cos\beta \ . \end{aligned} \tag{1.34}$$

Knowing both T and V in terms of the single variable w, the Lagrangian $L = T - V$ for the system is

$$L(w, \dot{w}) = m\dot{w}^2 - 2mgw\sin\beta + mgd\cos\beta. \tag{1.35}$$

The fact that this complicated system has only one degree of freedom might appear a bit surprising. The simplicity of the Lagrangian is a reflection of the symmetry of the system. As we will see later, symmetry plays an important role in solving problems in physics.

The Euler-Lagrange equation of the system becomes

$$2m\ddot{w} + 2mg\sin\beta = 0 \ . \tag{1.36}$$

Thus, the point P moves with a constant acceleration

$$\ddot{w} = -g\sin\beta = -g\sin(\alpha/2). \tag{1.37}$$

As one would expect, when $\alpha = 0$, the acceleration $\ddot{w}$ vanishes (which explains why we are able to hang our coats on nails), while for $\alpha \to \pi$ it becomes $-g$.

In the following example, we will consider a three-dimensional system with one constraint, and thus, only two degrees of freedom.

Example 1.5 The spherical pendulum

A spherical pendulum consists of a mass m on the end of an ideal string of length l as illustrated in Fig 1.4. The mass can move in any direction as long as the string remains taut. We are asked to describe the motion of mass m.

In spherical coordinates, the position of mass m is

$$\begin{aligned} x &= l\sin\theta\cos\varphi \\ y &= l\sin\theta\sin\varphi \\ z &= -l\cos\theta \end{aligned} \tag{1.38}$$

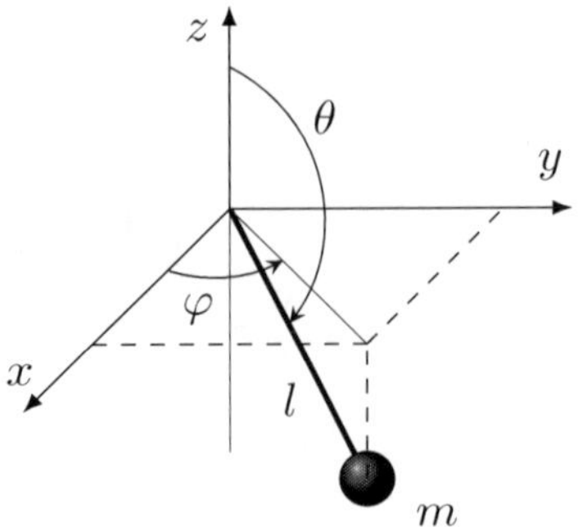

FIGURE 1.4
The spherical pendulum.

The kinetic and potential energies of the system are

$$T = \frac{1}{2}mv^2 = \frac{1}{2}m(\dot{x}^2 + \dot{y}^2 + \dot{z}^2) = \frac{1}{2}ml^2(\dot{\theta}^2 + \sin^2\theta\,\dot{\varphi}^2) \tag{1.39}$$

$$V = mgz = -mgl\cos\theta \tag{1.40}$$

where we considered the zero potential energy at $z = 0$. Then, the Lagrangian of the system is

$$L(\theta, \varphi, \dot{\theta}, \dot{\varphi}) = \frac{1}{2}ml^2(\dot{\theta}^2 + \sin^2\theta\,\dot{\varphi}^2) + mgl\cos\theta \tag{1.41}$$

Note that the Lagrangian is independent on φ. Therefore, the corresponding Lagrange equation reads

$$\frac{d}{dt}\left(\frac{\partial L}{\partial \dot{\varphi}}\right) = 0 , \tag{1.42}$$

which means that the generalized momentum

$$p_\varphi = \frac{\partial L}{\partial \dot{\varphi}} = ml^2\sin^2\theta\,\dot{\varphi} \tag{1.43}$$

is a constant of the motion. We recognize that p_ϕ is the angular momentum of the system about the z-axis. This angular momentum is conserved because the tension in the string does not exert a torque about the z-axis. Taking into account the conservation of angular momentum, we rewrite Eq. (1.43) as

$$\sin^2\theta\,\dot{\varphi} = B , \tag{1.44}$$

where B is a constant.

The second Lagrange equation

$$\frac{d}{dt}\left(\frac{\partial L}{\partial \dot{\theta}}\right) - \frac{\partial L}{\partial \theta} = 0 \tag{1.45}$$

becomes

$$ml^2\ddot{\theta} + mgl\sin\theta - ml^2\sin\theta\cos\theta\,\dot{\varphi}^2 = 0 \tag{1.46}$$

or

$$\ddot{\theta} + \frac{g}{l}\sin\theta - B^2\frac{\cos\theta}{\sin^3\theta} = 0\ , \tag{1.47}$$

where we took into account Eq. (1.44).

There are two important particular cases here. First, when the angular coordinate $\varphi = \varphi_0 =$ constant. Then from Eq. (1.44), $B = 0$, and Eq. (1.47) yields

$$\ddot{\theta} + \frac{g}{l}\sin\theta = 0\ . \tag{1.48}$$

This is exactly the equation of motion of the simple pendulum. Its motion is restricted to the vertical plane $\varphi = \varphi_0 =$ constant.

Second, we consider $\theta = \theta_0 =$ constant. In this case, from Eq. (1.44) we get $\dot{\varphi} = \dot{\varphi}_0 =$ constant, which implies that the pendulum bob rotates uniformly in the horizontal plane $z = l\cos\theta_0 \equiv z_0$. From Eqs. (1.44) and (1.47) we get

$$\dot{\varphi} \equiv \dot{\varphi}_0 = \sqrt{\frac{g}{z_0}}\ . \tag{1.49}$$

This type of pendulum is also called a *conical pendulum.*

This last example brought up a very important concept: *the constants of motion.* When the Lagrangian of a system doesn't depend explicitly on a given coordinate q_i, then $\partial L/\partial q_i = 0$, and consequently, from the Lagrange equation for coordinate q_i, we find that the corresponding generalized momentum

$$p_i = \frac{\partial L}{\partial \dot{q}_i} \tag{1.50}$$

is constant in time[1]. These constants are characteristic to the given system and are called *constants of motion*, and the corresponding coordinates *cyclic coordinates.* In our example we found that the angular momentum of the pendulum about the z-axis is a constant of the motion, and the corresponding cyclic coordinate was the azimuthal angle φ .

The Lagrangian formalism is concise and elegant. The entire information about the dynamics of the system is encoded into a single scalar function, the Lagrangian. Solving the Euler-Lagrange equations gives the complete description of the system's evolution, without the need of any force diagrams, depicting the interactions between its subsystems. The power and elegance of the Lagrangian formalism is the main reason that today, it has become a fundamental stepping stone for all areas of theoretical physics.

[1]Henceforth we will drop the ˜ from the momentum notation and write simply p instead of $\tilde{p}$.

1.4 Time invariance and the Hamiltonian function

In classical mechanics, for any inertial frame of reference, time is homogeneous. That means that the Lagrangian function for an isolated system cannot depend explicitly on time, or $\partial L/\partial t = 0$. Then, the time derivative of $L(q_i, \dot{q}_i)$ gives

$$\frac{dL}{dt} = \sum_{i=1}^{n} \frac{\partial L}{\partial q_i}\frac{dq_i}{dt} + \sum_{i=1}^{n} \frac{\partial L}{\partial \dot{q}_i}\frac{d\dot{q}_i}{dt} \tag{1.51}$$

$$\frac{dL}{dt} = \sum_{i=1}^{n} \frac{d}{dt}\left(\frac{\partial L}{\partial \dot{q}_i}\right)\frac{dq_i}{dt} + \sum_{i=1}^{n} \frac{\partial L}{\partial \dot{q}_i}\frac{d\dot{q}_i}{dt} \tag{1.52}$$

where in the second equation we made use of Lagrange equations $\frac{\partial L}{\partial q_i} = \frac{d}{dt}\left(\frac{\partial L}{\partial \dot{q}_i}\right)$. Taking into account that the generalized momentum associated with coordinate q_i is $p_i = \partial L/\partial \dot{q}_i$ we can rewrite Eq. (1.52) as

$$\frac{dL}{dt} = \sum_{i=1}^{n} \frac{d}{dt}\left(\frac{\partial L}{\partial \dot{q}_i}\dot{q}_i\right) = \sum_{i=1}^{n} \frac{d}{dt}\left(p_i\dot{q}_i\right) \tag{1.53}$$

or

$$\frac{d}{dt}\left(\sum_{i=1}^{n} p_i\dot{q}_i - L\right) = 0 \ . \tag{1.54}$$

Thus, the quantity in parentheses is constant in time. This constant of motion, denoted by H, is called the Hamiltonian function.

$$H = \sum_{i=1}^{n} p_i\dot{q}_i - L = \text{constant.} \tag{1.55}$$

It is important to note that H is a constant of the motion only if the Lagrangian function doesn't depend explicitly on time.

One might wonder why in Eq. (1.54) we chose that particular order of terms and not the reverse one. The reason will become clear shortly. Let us consider a holonomic system which is subject to scleronomic constraints; i.e., the equations of the constraints do not contain the time as an explicit variable:

$$x_k^\ell = x_k^\ell(q_1, q_2, \ldots, q_n) \quad \text{where } k = 1, 2, \ldots N; \quad \ell = 1, 2, 3 \ . \tag{1.56}$$

With these assumptions, we will show that the kinetic energy remains a homogeneous quadratic function of the generalized velocities. First, note that in

Cartesian coordinates, the kinetic energy is already a homogeneous quadratic function:

$$T = \frac{1}{2}\sum_{k=1}^{N}\sum_{\ell=1}^{3} m_k \left(\dot{x}_k^\ell\right)^2 \ , \tag{1.57}$$

where m_k is the mass of the k-th particle and $\dot{x}_k^\ell$ is the ℓ-th component of its velocity. In generalized coordinates q_i, the kinetic energy becomes

$$\begin{aligned} T &= \frac{1}{2}\sum_{k=1}^{N}\sum_{\ell=1}^{3} m_k \left(\frac{d\,x_k^\ell}{dt}\right)^2 = \frac{1}{2}\sum_{k=1}^{N}\sum_{\ell=1}^{3} m_k \left(\sum_{i=1}^{n}\frac{\partial\,x_k^\ell}{\partial q_i}\frac{dq_i}{dt}\right)\left(\sum_{j=1}^{n}\frac{\partial\,x_k^\ell}{\partial q_j}\frac{dq_j}{dt}\right) \\ &= \frac{1}{2}\sum_{i,j=1}^{n}\left(\sum_{k=1}^{N}\sum_{\ell=1}^{3} m_k \frac{\partial\,x_k^\ell}{\partial q_i}\,\frac{\partial\,x_k^\ell}{\partial q_j}\right)\frac{dq_i}{dt}\,\frac{dq_j}{dt} \ . \end{aligned} \tag{1.58}$$

Therefore

$$T = \frac{1}{2}\sum_{i,j=1}^{n} T_{ij}(q)\,\dot{q}_i\dot{q}_j \ , \tag{1.59}$$

where $T_{ij}(q) = \sum_{k=1}^{N}\sum_{\ell=1}^{3} m_k \frac{\partial\,x_k^\ell}{\partial q_i}\,\frac{\partial\,x_k^\ell}{\partial q_j}$. Thus the kinetic energy T is a quadratic homogeneous function of the generalized velocities $\dot{q}_i$. Therefore we have

$$\sum_{k=1}^{n}\dot{q}_k\frac{\partial T}{\partial \dot{q}_k} = \frac{1}{2}\sum_{k=1}^{n}\dot{q}_k\sum_{i,j=1}^{n} T_{ij}\left(\delta_{ik}\dot{q}_j + \delta_{jk}\dot{q}_i\right) = \sum_{k,j=1}^{n} T_{kj}\dot{q}_k\dot{q}_j = 2T, \tag{1.60}$$

which is a particular case of Euler's theorem[2] for homogeneous functions.

If the potential energy does not depend on the generalized velocities $\dot{q}_i$ then,

$$\frac{\partial L}{\partial \dot{q}_i} = \frac{\partial (T-V)}{\partial \dot{q}_i} = \frac{\partial T}{\partial \dot{q}_i} \ . \tag{1.61}$$

Combining Eqs. (1.55) with (1.60) and (1.61), we obtain

$$H = 2T - (T-V) = T + V = E = \text{constant}, \tag{1.62}$$

where E is the total mechanical energy of the system. Now it is clear the reason for the sign choice in Eq. (1.54). Using this choice, for a large class of systems, the Hamiltonian function is equal to the total energy of the system.

It is important to remember that the Hamiltonian function is not always equal to the total energy of the system. This is true only for holonomic systems subject to scleronomic constraints, and if the potential energy doesn't depend on the velocities.

---

[2]For a homogeneous function $f(q_1, q_2, \ldots, q_n)$ of order M in variables $(q_1, q_2, \ldots, q_n)$, we have $\sum_{j=1}^{n} q_j(\partial f/\partial q_j) = Mf$.

1.5 Hamilton's equations of motion

In general, the Lagrangian L function depends on the generalized coordinates, the generalized velocities, and it may be an explicit function of time

$$L = L(q_1, \ldots, q_n; \dot{q}_1, \ldots, \dot{q}_n; t) \ . \tag{1.63}$$

On the other hand, the Hamiltonian function H is defined as

$$H = \sum_{i=1}^{n} p_i \dot{q}_i - L(q_1, \ldots, q_n; \dot{q}_1, \ldots, \dot{q}_n; t) \ , \tag{1.64}$$

where the generalized momenta p_i are given by $p_i = \partial L / \partial \dot{q}_i$. The generalized velocities $\dot{q}_i$ are to be expressed as functions of q_i and p_i only. With this, the Hamiltonian becomes a function only of the generalized coordinates, generalized momenta, and time

$$H = H(q_1, \ldots, q_n; p_1, \ldots, p_n; t) \ . \tag{1.65}$$

Then,

$$dH = \sum_{i=1}^{n} \frac{\partial H}{\partial q_i} dq_i + \sum_{i=1}^{n} \frac{\partial H}{\partial p_i} dp_i + \frac{\partial H}{\partial t} dt \tag{1.66}$$

From Eq. (1.64), we write the total differential of the Hamiltonian as

$$dH = \sum_{i=1}^{n} p_i d\dot{q}_i + \sum_{i=1}^{n} \dot{q}_i dp_i - dL \tag{1.67}$$

Note that using Lagrange equations $\dot{p}_i = \partial L / \partial q_i$, we can write the differential of the Lagrangian as

$$\begin{aligned} dL &= \sum_{i=1}^{n} \frac{\partial L}{\partial q_i} dq_i + \sum_{i=1}^{n} \frac{\partial L}{\partial \dot{q}_i} d\dot{q}_i + \frac{\partial L}{\partial t} dt \\ &= \sum_{i=1}^{n} \dot{p}_i dq_i + \sum_{i=1}^{n} p_i d\dot{q}_i + \frac{\partial L}{\partial t} dt \end{aligned} \tag{1.68}$$

which, substituted back into Eq. (1.67) gives

$$dH = -\sum_{i=1}^{n} \dot{p}_i dq_i + \sum_{i=1}^{n} \dot{q}_i dp_i - \frac{\partial L}{\partial t} dt \ . \tag{1.69}$$

Comparing Eqs (1.66) and(1.69), we obtain Hamilton equations:

$$\dot{q}_i = \frac{\partial H}{\partial p_i} \ , \qquad \dot{p}_i = -\frac{\partial H}{\partial q_i} \ , \qquad \frac{\partial H}{\partial t} = -\frac{\partial L}{\partial t} \ . \tag{1.70}$$

From Eq. (1.69) and the third Hamilton's equation (1.70), we obtain

$$\frac{dH}{dt} = \frac{\partial H}{\partial t} , \tag{1.71}$$

which implies that if the Hamiltonian is not an explicit function of time then it is conserved.

Note that while the Lagrange equations provide a set of n second order differential equations for the generalized coordinates, the Hamilton equations provide a set of $2n$ first order differential equations for the coordinates and momenta. In both, cases we need $2n$ initial conditions when solving the system of equations.

Thus, to apply the Hamiltonian formalism we follow the steps below:

1. Find $L(q_i, \dot{q}_i, t)$,
2. Write $p_i = \partial L(q_i, \dot{q}_i, t)/\partial \dot{q}_i$,
3. Solve for the generalized velocities $\dot{q}_i = \dot{q}_i(q_i, p_i)$,
4. Substitute the $\dot{q}_i$ into $H = \sum p_i q_i - L(q_i, \dot{q}_i, t)$, which now becomes a function only of q_i and p_i,
5. Write Hamilton's equations.

Example 1.6 A diatomic molecule

We can model a diatomic molecule as a system comprised of two particles of masses m_1 and m_2 connected by a spring of constant k, as shown in Fig. 1.5. Let us derive the Hamilton's equations of motion for the two masses.

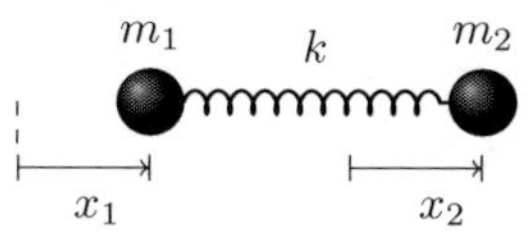

FIGURE 1.5
Two particles of masses m_1 and m_2 connected by a spring of constant k.

We choose the Cartesian system of coordinates and denote the positions of the two balls from their equilibrium points by x_1 and x_2. Then, the Lagrangian of the system is

$$L(x_1, x_2, \dot{x}_1, \dot{x}_2) = T - V = \frac{1}{2}m_1\dot{x}_1^2 + \frac{1}{2}m_2\dot{x}_2^2 - \frac{1}{2}k(x_1 - x_2)^2 . \tag{1.72}$$

The generalized momenta are

$$p_1 \equiv \frac{\partial L}{\partial \dot{x}_1} = m_1\dot{x}_1 \ ; \quad p_2 \equiv \frac{\partial L}{\partial \dot{x}_2} = m_2\dot{x}_2 . \tag{1.73}$$

In terms of the generalized momenta, the generalized velocities are given by

$$\dot{x}_1 = \frac{p_1}{m_1} \ ; \quad \dot{x}_2 = \frac{p_1}{m_2} . \tag{1.74}$$

Therefore, the Hamiltonian $H = T + V$ of the system becomes

$$H = \frac{p_1^2}{2m_1} + \frac{p_2^2}{2m_2} + \frac{1}{2}k(x_1 - x_2)^2 \,, \tag{1.75}$$

which gives the following set of Hamilton's equations

$$\dot{x}_1 = \frac{\partial H}{\partial p_1} = \frac{p_1}{m} \,; \quad \dot{p}_1 = -\frac{\partial H}{\partial x_1} = -k(x_1 - x_2) \,, \tag{1.76}$$

$$\dot{x}_2 = \frac{\partial H}{\partial p_2} = \frac{p_2}{m} \,; \quad \dot{p}_2 = -\frac{\partial H}{\partial x_2} = -k(x_2 - x_1) \,, \tag{1.77}$$

which is what we would have expected from applying the usual Newtonian mechanics to the system. Along these lines, why do you think we have $\dot{p}_1 = -\dot{p}_2$ in the equations above?

Exercise 1.2 Derive the Hamiltonian of Eq. (1.75) starting from the general definition (1.64).

The next example is a bit more complicated.

Example 1.7 The double pendulum revisited
Let us consider the example of the double pendulum from the previous Example 1.3, which is described by the Lagrangian

$$L(\theta_1, \theta_2, \dot{\theta}_1, \dot{\theta}_2) = \frac{1}{2}(m_1 + m_2)\, l_1^2\, \dot{\theta}_1^{\,2} + \frac{1}{2} m_2\, l_2^2\, \dot{\theta}_2^{\,2} + \dot{\theta}_1 \dot{\theta}_2\, m_2\, l_1 l_2 \cos(\theta_1 - \theta_2)$$
$$+(m_1 + m_2)\, g l_1 \cos\theta_1 + m_2\, g\, l_2 \cos\theta_2 \tag{1.78}$$

The generalized momenta associated to the generalized coordinates θ_1 and θ_2 are

$$p_{\theta_1} \equiv \frac{\partial L}{\partial \dot{\theta}_1} = (m_1 + m_2)\, l_1^2 \dot{\theta}_1 + \dot{\theta}_2\, m_2\, l_1 l_2 \cos(\theta_1 - \theta_2)$$
$$p_{\theta_2} \equiv \frac{\partial L}{\partial \dot{\theta}_2} = m_2\, l_2^2 \dot{\theta}_2 + \dot{\theta}_1\, m_2\, l_1 l_2 \cos(\theta_1 - \theta_2) \tag{1.79}$$

Note that these generalized momenta do not have the dimension of the linear momentum. However, the product of the generalized coordinate and the generalized momentum always has the dimension of angular momentum.

Because the potential energy doesn't depend on velocities, the equations (1.79) are linear in $\dot{\theta}_i$ and can be immediately solved in terms of θ_i and p_{θ_i}

$$\begin{aligned} \dot{\theta}_1 &= p_{\theta_1}\, m_2 l_2^2\, A + p_{\theta_2} m_2 l_1 l_2 \cos(\theta_1 - \theta_2)\, A \\ \dot{\theta}_2 &= -p_{\theta_1} m_2 l_1 l_2 \cos(\theta_1 - \theta_2)\, A + p_{\theta_2}(m_1 + m_2) l_1^2\, A \end{aligned} \tag{1.80}$$

where we denoted

$$A = \left[{l_1}^2{l_2}^2 m_2 \left(m_1 + m_2 - m_2 \cos^2(\theta_1 - \theta_2)\right)\right]^{-1} .$$

Note that the expressions of $\dot{\theta}_i$ are quite complicated, which in turn yield a prohibitively complicated Hamiltonian function.

$$\begin{aligned} H &= \frac{A^2}{2}(m_1 + m_2)\, l_1^2 \,(p_{\theta_1}\, m_2 l_2^2 + p_{\theta_2} m_2 l_1 l_2 \cos(\theta_1 - \theta_2))^2 \\ &+ \frac{A^2}{2} m_2\, l_2^2 \,(-p_{\theta_1} m_2 l_1 l_2 \cos(\theta_1 - \theta_2) + p_{\theta_2}(m_1 + m_2) l_1^2)^2 \\ &+ A^2 (p_{\theta_1}\, m_2 l_2^2 + p_{\theta_2} m_2 l_1 l_2 \cos(\theta_1 - \theta_2))\, m_2\, l_1 l_2 \cos(\theta_1 - \theta_2) \\ &\quad \times(-p_{\theta_1} m_2 l_1 l_2 \cos(\theta_1 - \theta_2) + p_{\theta_2}(m_1 + m_2) l_1^2) \\ &- (m_1 + m_2)\, g l_1 \cos\theta_1 - m_2\, g\, l_2 \cos\theta_2 \end{aligned} \tag{1.81}$$

Consequently, the associated Hamilton's equations become mixed first order transcendental differential equations, which are hard to solve. In the next chapter we will see that in the approximation of small oscillations, the analytic description of the double pendulum simplifies considerably.

Exercise 1.3 Derive equations (1.80).

Example 1.8 Bead on a rotating parabolic wire

Let us consider the case of a vertical parabolic wire, described in cylindrical coordinates by $z = br^2$, where r is the distance of a point on the wire to the vertical axis z which is also the symmetry axis of the parabola. If the wire rotates uniformly with angular velocity ω about z, we want to determine the motion of a bead of mass m which can slide without friction along the wire.

In the first step we find the number of degrees of freedom. The bead moves in three dimensions, but it is constrained to slide along the wire. Therefore, if we know the distance r, we also know its vertical coordinate z because $z = br^2$. Further, we notice that the azimuthal angle φ is constrained by the uniform rotation of the wire $\varphi = \varphi_0 + \omega t$, where ω is constant. We have three coordinates, and two constraints, which leaves us with only one degree of freedom. Therefore, we could choose the generalized coordinate to be the distance r.

The square of the bead's velocity, in cylindrical coordinates (r, φ, z) is then

$$v^2 = v_r^2 + v_\varphi^2 + v_z^2 = \dot{r}^2 + r^2\dot{\varphi}^2 + \dot{z}^2 = \dot{r}^2 + r^2\omega^2 + (2br\dot{r})^2 \;, \tag{1.82}$$

where we used the constraints $z = br^2$ and $\dot{\varphi} = \omega$ =constant. The potential energy is the gravitational potential energy $V = mgz = mgb\,r^2$. Then, the Lagrangian of the system is

$$L(r, \dot{r}) = T - V = \frac{m}{2}\left[\omega^2 r^2 + (1 + 4b^2r^2)\,\dot{r}^2\right] - mgb\,r^2 \;. \tag{1.83}$$

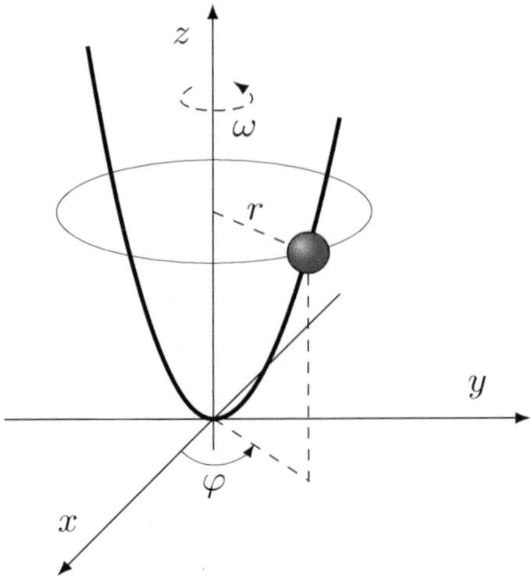

FIGURE 1.6
A bead of mass m sliding without friction on a vertical parabolic wire that rotates uniformly about the z-axis with angular velocity ω.

The Lagrangian function is not an explicit function of time, $\partial L/\partial t = 0$, therefore, from the third Hamilton's equation (1.70), we obtain that $\partial H/\partial t = 0$ and hence $dH/dt = 0$. So the Hamiltonian is a conserved quantity, which will play a significant role going forward. To find the Hamiltonian, we first calculate the generalized momentum

$$p_r = \frac{\partial L}{\partial \dot{r}} = m\,(1 + 4b^2\,r^2)\,\dot{r}\;. \tag{1.84}$$

Then, the value of the Hamiltonian is

$$\begin{aligned} H &= \dot{r}p_r - L = m\,(1+4b^2\,r^2)\,\dot{r}^2 - \frac{m}{2}\left[\omega^2 r^2 + (1+4b^2 r^2)\,\dot{r}^2\right] + mgb\,r^2 \\ &= \frac{m}{2}\left[(1+4b^2\,r^2)\,\dot{r}^2 - \omega^2 r^2\right] + mgb\,r^2\;. \end{aligned}$$

Expressing $\dot{r}$ in terms of p_r and r using Eq. (1.84), we obtain finally

$$H = \frac{p_r^2}{2m(1+4b^2\,r^2)} + \frac{m}{2}(2gb - \omega^2)r^2 = \text{constant}. \tag{1.85}$$

Note that the total energy of the bead becomes

$$E = T + V = \frac{p_r^2}{2m(1+4b^2\,r^2)} + \frac{m}{2}(2gb + \omega^2)r^2\;, \tag{1.86}$$

which is different from the Hamiltonian. Their difference is

$$E - H = m\omega^2 r^2 \tag{1.87}$$

which changes with r. It is important to observe that while the Hamiltonian H is conserved, the energy E is not, because the normal force of the wire does work on the bead. In this case, the normal force is not perpendicular to the displacement. This work is associated with the energy input imparted by the rotating wire to the bead.

Because the Hamiltonian is constant, from Eq. (1.85) we obtain a first order differential equation which can be further solved. We will not solve the equation, but rather show how the particular form of the Hamiltonian encodes a trove of information about the motion of the bead. For this, let us rewrite Eq. (1.85) as

$$H = \frac{p_r^2}{2m(1+4b^2\, r^2)} + \widetilde{V}(r) \; , \tag{1.88}$$

where

$$\widetilde{V}(r) = \frac{m}{2}(2bg - \omega^2)\, r^2 \tag{1.89}$$

plays the role of an effective quadratic potential. The behavior of the bead on the rotating wire depends on the sign of the coefficient of r^2 which in turn depends on the relation between the angular velocity ω and the critical angular velocity $\widetilde{\omega} = (2gb)^{1/2}$. If $\omega < \widetilde{\omega}$ then the effective potential increases with r, and consequently the bead is in a stable equilibrium at $r = 0$. See Fig. 1.7. If $\omega > \widetilde{\omega}$ then the effective potential decreases with r and $r = 0$ is an

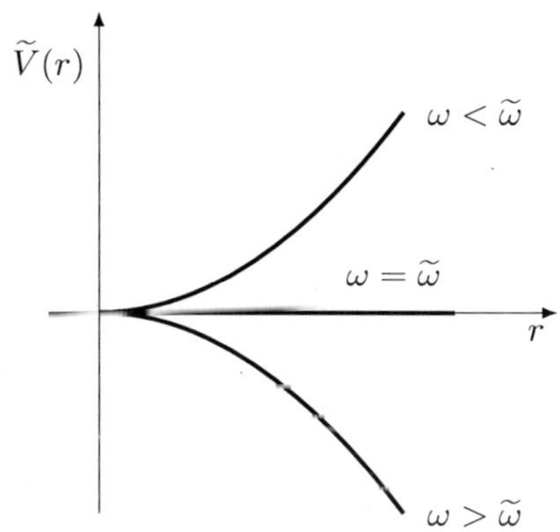

FIGURE 1.7
The effective potential for a bead on a rotating parabolic wire.

unstable equilibrium point. The slightest deviation will push the bead farther and farther away from the origin. For the particular case when $\omega = \widetilde{\omega}$ the bead will stay indefinitely at rest if initially placed at rest on any point on the wire.

Exercise 1.4 Prove that for $\omega = \widetilde{\omega}$ the bead will stay indefinitely at rest if initially placed at rest.

1.6 Total energy vs. Hamiltonian

In this last example we encountered a Hamiltonian H which was not equal to the total energy of the system E. The Hamiltonian was constant in time, while the energy E was not. The energy was not conserved in this case because the wire was continuously doing work on the bead, thus injecting energy into the system. As you may have already suspected, such situations occur when not all the constraints of the system are independent on time. That is exactly the case when the system is subject to rheonomic constraints. Now, we will prove this observation in general.

Let us consider a holonomic system which is subject to rheonomic constraints; i.e., the equations of the constraints as described in generalized coordinates contain the time as an explicit variable:

$$x_k^\ell = x_k^\ell(q_1, q_2, \ldots, q_n; t) \ , \quad \text{where } k = 1, 2, \ldots N; \quad \ell = 1, 2, 3 \ . \tag{1.90}$$

With these assumptions, we will return to the calculation of the kinetic energy when expressed in the generalized coordinates.

First, in Cartesian coordinates, the kinetic energy is:

$$T = \frac{1}{2} \sum_{k=1}^{N} \sum_{\ell=1}^{3} m_k \left(\dot{x}_k^\ell\right)^2 \quad , \tag{1.91}$$

where, as before, m_k is the mass of the k-th particle and $\dot{x}_k^\ell$ is the ℓ-th component of its velocity.

In generalized coordinates q_i, the kinetic energy becomes

$$\begin{aligned}
T &= \frac{1}{2} \sum_{k=1}^{N} \sum_{\ell=1}^{3} m_k \left(\frac{d\,x_k^\ell}{dt}\right)^2 \\
&= \frac{1}{2} \sum_{k=1}^{N} \sum_{\ell=1}^{3} m_k \left(\sum_{i=1}^{n} \frac{\partial\, x_k^\ell}{\partial q_i} \frac{dq_i}{dt} + \frac{\partial\, x_k^\ell}{\partial t}\right) \left(\sum_{j=1}^{n} \frac{\partial\, x_k^\ell}{\partial q_j} \frac{dq_j}{dt} + \frac{\partial\, x_k^\ell}{\partial t}\right) \\
&= \frac{1}{2} \sum_{i,j=1}^{n} T_{ij}(q)\, \dot{q}_i \dot{q}_j + \frac{1}{2} \sum_{k=1}^{N} \sum_{\ell=1}^{3} m_k \left(\frac{\partial\, x_k^\ell}{\partial t}\right)^2 + \sum_{k=1}^{N} \sum_{\ell=1}^{3} m_k \sum_{j=1}^{n} \frac{\partial\, x_k^\ell}{\partial q_i} \dot{q}_j \frac{\partial\, x_k^\ell}{\partial t}
\end{aligned}$$

where $T_{ij}(q) = \sum_{k=1}^{N} \sum_{\ell=1}^{3} m_k \frac{\partial\, x_k^\ell}{\partial q_i} \frac{\partial\, x_k^\ell}{\partial q_j}$. Consequently

$$\frac{\partial T}{\partial \dot{q}_j} = \sum_{i=1}^{n} T_{ij}(q)\, \dot{q}_i + \sum_{k=1}^{N} \sum_{\ell=1}^{3} m_k \frac{\partial\, x_k^\ell}{\partial q_i} \frac{\partial\, x_k^\ell}{\partial t} \ . \tag{1.92}$$

We thus obtain

$$\begin{aligned}
\sum_{j=1}^{n} \dot{q}_j p_j \equiv \sum_{j=1}^{n} \dot{q}_j \frac{\partial T}{\partial \dot{q}_j} &= \sum_{i,j=1}^{n} T_{ij}\dot{q}_j\dot{q}_i + \sum_{k=1}^{N}\sum_{\ell=1}^{3} m_k \sum_{j=1}^{n} \frac{\partial x_k^\ell}{\partial q_i}\dot{q}_j \frac{\partial x_k^\ell}{\partial t} \\
&= 2T - \left[\sum_{k=1}^{N}\sum_{\ell=1}^{3} m_k \left(\sum_{j=1}^{n} \frac{\partial x_k^\ell}{\partial q_i}\dot{q}_j + \frac{\partial x_k^\ell}{\partial t}\right)\frac{\partial x_k^\ell}{\partial t}\right] \\
&= 2T - \sum_{k=1}^{N}\sum_{\ell=1}^{3} m_k \frac{dx_k^\ell}{dt}\frac{\partial x_k^\ell}{\partial t} \qquad (1.93)
\end{aligned}$$

which is a generalization of Eq. 1.60. Therefore, in general, the Hamiltonian can be written as

$$H = \sum_{j=1}^{n} \dot{q}_j p_j - L = E - \sum_{k=1}^{N}\sum_{\ell=1}^{3} m_k \frac{dx_k^\ell}{dt}\frac{\partial x_k^\ell}{\partial t} \qquad (1.94)$$

where $E = T + V$ is the energy of the system. We see that if the constraints do not depend explicitly of time (scleronomic constraints), then $\frac{\partial x_k^\ell}{\partial t} = 0$, and the Hamiltonian is just $T + V$. But if the constraints are moving, then this is no longer the case, as we have seen in the case of the bead sliding on the rotating parabolic wire.

1.7 Problems

1.1 Find the number of degrees of freedom for a particle constrained to move on a (i) cone, (ii) cylinder, and (iii) paraboloid.

1.2 Find the Lagrangian function for a simple pendulum and write the corresponding Lagrange equations.

1.3 Use the Lagrange formalism to find the differential equations of motion for the physical pendulum of mass M and moment of inertia with respect to the center of mass I_{CM} which oscillates in a vertical plane about a fixed horizontal axis, as illustrated in Fig. 1.8.

1.4 A particle of mass m moves in the xy-plane under the influence of a force derived from a potential that depends only on the distance from the origin. Find the Lagrangian function and write the equations of motion.

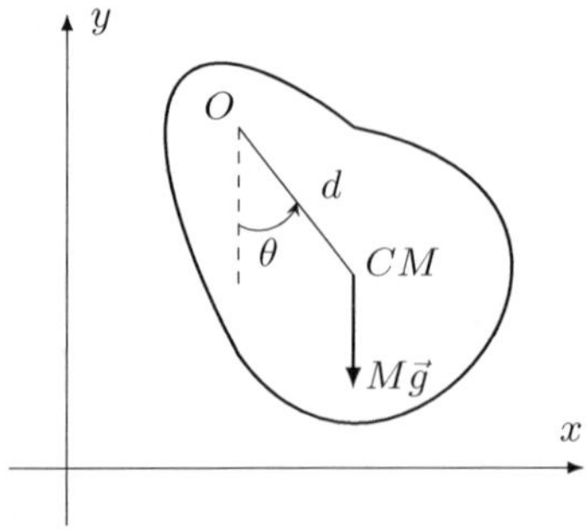

FIGURE 1.8
Physical pendulum for Problem 1.3.

1.5 The Lagrangian of a particle moving in three dimensions is given by $L = \frac{m}{2}(\dot{x}^2 + \dot{y}^2 + \dot{z}^2) + A\dot{y} + B\dot{z} + C$, where A, B, and C are constants.

(a) Write Lagrange equations of motion and show that the particle moves uniformly in a straight line.

(b) This result hints at the existence of another frame of reference (X, Y, Z) in which the Lagrangian is exactly the Lagrangian of a free particle $L' = \frac{m}{2}(\dot{X}^2 + \dot{Y}^2 + \dot{Z}^2)$. Find the new set of coordinates (X, Y, Z) in terms of (x, y, z) and its velocity with respect to the initial frame of reference.

1.6 A particle of mass m moves without friction under the action of gravity on the inner surface of a paraboloid of revolution described by $x^2 + y^2 = \lambda z$ as illustrated in Fig. 1.9. The constant λ has dimensions of length. Use the Lagrangian formalism to find the equations of motion.

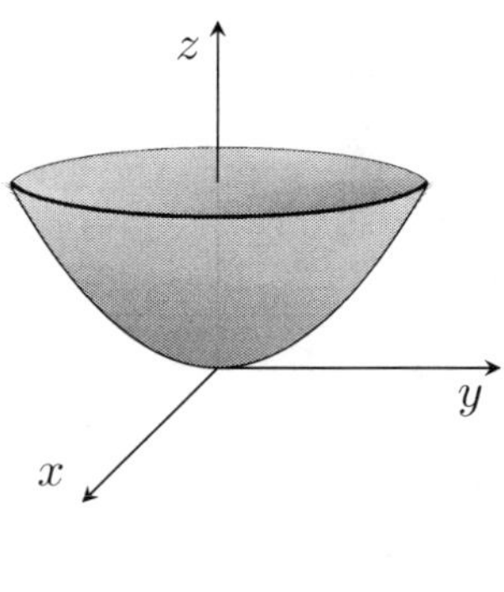

FIGURE 1.9
The paraboloidal surface for the motion of particle in Problem 1.6.

1.7 A mass m glides frictionless along a massless rod which can oscillate only in the vertical plane. The mass is connected to the rod's joint via a massless spring of elastic constant k which coils around the rod as illustrated in Fig. 1.10. Write the Lagrange equations and discuss the case of the oscillations with very small angles, when we can approximate $\sin\theta \approx \theta$.

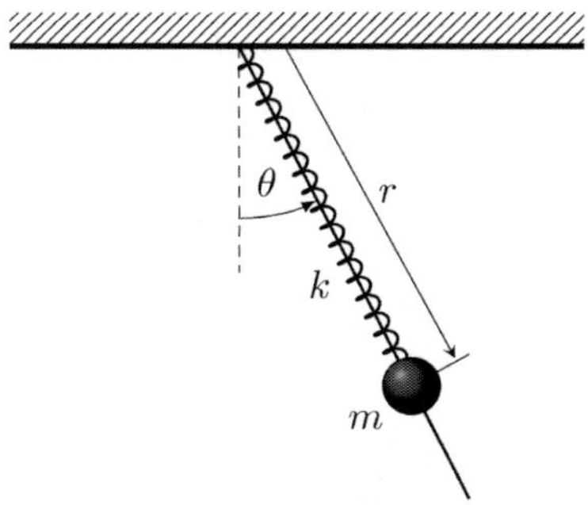

FIGURE 1.10
Elastic oscillating rod for Problem 1.7.

1.8 The Hamiltonian of two particles of masses m_1 and m_2 is given by $H(\vec{r}_1, \vec{r}_2, \vec{p}_1, \vec{p}_2) = \frac{1}{2m_1}|\vec{p}_1|^2 + \frac{1}{2m_2}|\vec{p}_2|^2 + V(\vec{r}_1 - \vec{r}_2)$.

(a) Rewrite the Hamiltonian in the center of mass coordinates given by $\vec{R} = \frac{m_1\vec{r}_1 + m_2\vec{r}_2}{m_1+m_2}$ and relative coordinates $\vec{r} = \vec{r}_1 - \vec{r}_2$.

(b) Show that the differential equations of motion decouple into equations describing the uniform translation of the center of mass and equations describing the motion of a reduced mass particle $\mu = \frac{m_1 m_2}{m_1+m_2}$ in the effective central potential $V(\vec{r})$.

1.9 Two Lagrangians L and L' are called equivalent if they differ by the total time derivative of some function $\Lambda(q,t)$, i.e.; $L' = L + d\Lambda/dt$.

(a) Prove that two equivalent Lagrangians yield the same Lagrange equations.

(b) Find the connection between the generalized momenta p and p' generated by the equivalent Lagrangians L and L'.

(c) Find the relation between H and H'.

1.10 The Lagrangian of a particle of mass m moving in a force field generated by the potential $V(r)$ is given in spherical coordinates by $L(r, \theta, \varphi, \dot{r}, \dot{\theta}, \dot{\varphi}) = \frac{1}{2}m\left(\dot{r}^2 + r^2\dot{\theta}^2 + r^2\sin^2\theta\,\dot{\varphi}^2\right) - V(r)$.

(a) Calculate the generalized momenta $(p_r, p_\theta, p_\varphi)$ conjugate to (r, θ, φ).

(b) Find the Hamiltonian $H(r, \theta, \varphi, p_r, p_\theta, p_\varphi)$ and write the corresponding Hamilton equations.

1.11 A rigid wire is bent in a shape of a **V** with the middle angle of 2β. The V-shaped wire is then held vertically and uniformly rotated with the

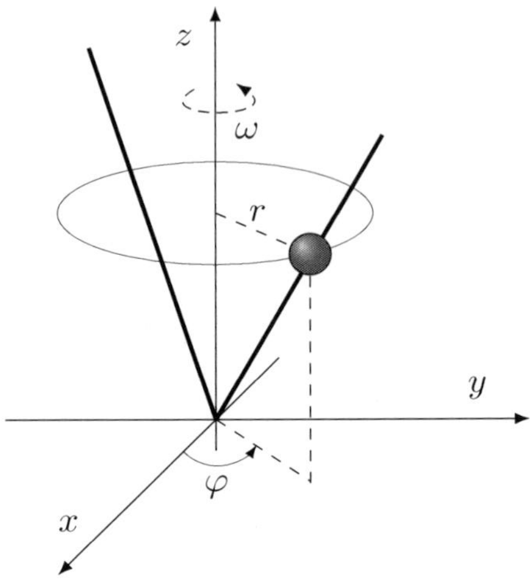

FIGURE 1.11
Figure for Problem 1.11: a bead of mass m sliding without friction on a V-shaped vertical wire that rotates uniformly about the z-axis with angular velocity ω.

angular velocity ω along a vertical axis that passes through the bisector of the vertex angle. A bead of mass m is inserted on one of the arms of the rotating V-shaped wire.

(a) Identify explicitly the constraints of the system and define a generalized coordinate for the bead.

(b) Write the Lagrangian and calculate the generalized momentum

(c) Find the Hamiltonian. Is it conserved?

(d) Find the effective potential of the system and identify and describe any points of equilibrium.

1.8 Appendix: finding the dynamical angle

Here we return to example 1.4 and prove that the dynamical angle β is $\alpha/2$ for the horizontal table. We will consider that the table is tilted by angle θ

symmetrically about the bisector of angle α, as shown in Fig. 1.12 Moreover,

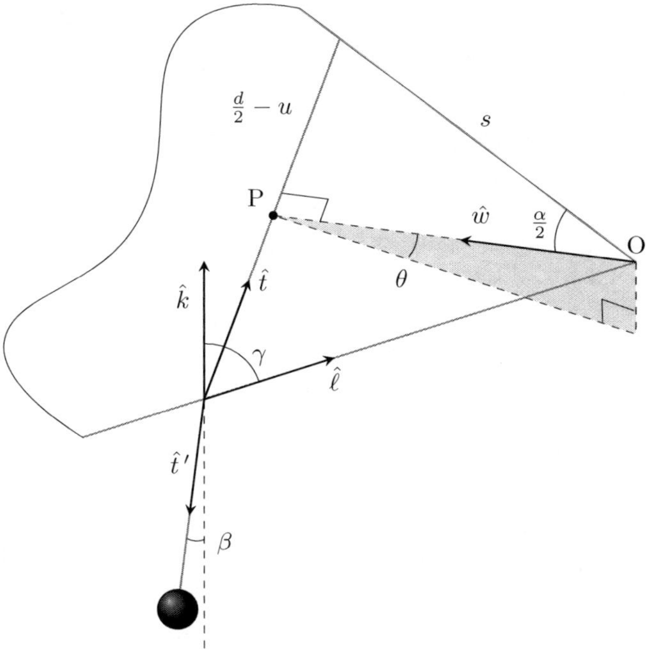

FIGURE 1.12
Illustration of the dynamical constraints in the general case of the table tilted by and angle θ.

we assume the string is oriented so that the portion in contact with the top of the table is always horizontal. Because the table is assumed frictionless, the interaction between a piece of string and the table will always be perpendicular to the surface of the table, we denoted by $\vec{N}$ the force of the table on the small portion of the string that is in contact with the edge of the table. The tension forces acting on this segment of string are given by $T\hat{t}$ for the portion of the string on the table, and by $T\hat{t}'$ for the hanging portion of the string, where $\hat{t}$ and $\hat{t}'$ are unit vectors that point along these string segments (see Fig. 1.12). Note that, owing to our assumptions of a massless string and the smoothness of the table, the tension can be safely assumed to be uniform along the string. Because the string is strictly massless, the net force on any segment must be zero; applying this to the portion of the string in contact with the edge of the table then gives $\vec{N} + T\hat{t} + T\hat{t}' = 0$. This equation tells us that $\hat{t} + \hat{t}' = -\vec{N}/T$, and this vector must be perpendicular to the unit vector $\hat{\ell}$ that points along the edge of the table.

Considering that the motion of mass A is confined to a vertical plane containing the edge of the table, we see from Fig. 1.12 that we can write the unit vector $\hat{t}'$ as a linear combination of the unit vectors $\hat{\ell}$ and $\hat{z}$

$$\hat{t}' = C_1\hat{\ell} + C_2\hat{z} \tag{1.95}$$

where C_1 and C_2 are constant coefficients. We need to determine these coefficients in order to find the angle β. Since $\hat{t}'$ is a unit vector we have

$$\hat{t}' \cdot \hat{t}' = 1 = C_1^2 + C_2^2 + 2C_1C_2 \cos\gamma \ , \tag{1.96}$$

where $\cos\gamma = \hat{\ell} \cdot \hat{z}$. Moreover, because $\hat{t} \cdot \hat{\ell} = \sin(\alpha/2)$, the fact $\hat{t} + \hat{t}'$ is perpendicular to ℓ implies that $\hat{t}' \cdot \hat{\ell} = -\hat{t} \cdot \hat{\ell} = -\sin(\alpha/2)$. Hence, taking the dot product of $\hat{\ell}$ with Eq. (1.95), we find that

$$C_1 + C_2 \cos\gamma = -\sin\left(\frac{\alpha}{2}\right) \ . \tag{1.97}$$

Solving Eqs. (1.96) and (1.97) for C_2, we find

$$C_2 = \pm\frac{cos(\alpha/3)}{\sin\gamma} \tag{1.98}$$

and, guided by Fig. 1.12, we choose the negative sign. Having found C_2, the coefficient C_1 is then found to be

$$C_1 = -C_2 \cos\gamma - \sin\left(\frac{\alpha}{2}\right) = \frac{cos(\alpha/2 + \gamma)}{\sin\gamma} \ . \tag{1.99}$$

We can now determine the angle β defined by $\cos\beta = \hat{t}'\cdot(-\hat{z})$. This angle is a crucial parameter that is related to the acceleration in the forward direction because the masses only accelerate horizontally when $\beta \neq 0$. Using Eq. (1.95) we have

$$\begin{aligned} \cos\beta &= \hat{t}' \cdot (-\hat{z}) = -C_1 \cos\gamma - C_2 \\ &= -\frac{\cos\gamma}{\sin\gamma} \cos\left(\frac{\alpha}{2} + \gamma\right) + \frac{\cos(\alpha/2)}{\sin\gamma} \\ &= \sin\left(\frac{\alpha}{2} + \gamma\right) \\ &= \cos\left(\frac{\pi}{2} - \frac{\alpha}{2} - \gamma\right) \end{aligned} \tag{1.100}$$

which gives

$$\beta = \pm\left(\frac{\pi}{2} - \frac{\alpha}{2} - \gamma\right) \tag{1.101}$$

Note that for the horizontal table, with $\theta = 0$ and $\gamma = \pi/2$, the angel β should be positive. Thus, choosing the negative sign in Eq. (1.101), we obtain

$$\beta = \frac{\alpha}{2} + \gamma - \frac{\pi}{2} \ , \tag{1.102}$$

which for the horizontal table gives

$$\beta = \frac{\alpha}{2} \,. \tag{1.103}$$

Note that if we start the system from a different initial configuration where β differs from $\alpha/2$ and the two masses hang symmetrically at a given depth below the table, then the angle β will not remain constant, and the onset of the motion will induce oscillations of the moving masses.

2

Small Oscillations

2.1 Introduction

Oscillatory motions are ubiquitous in our everyday life. A vibrating refrigerator, a Grand-Father clock, plucked strings of a guitar, they all are examples of oscillatory motions. Oscillations are also found at all scales, from molecular level to astronomical objects.

In this chapter we will study small oscillations, i.e., oscillations with amplitudes that are small enough that the restoring forces can be approximated to be linearly proportional to the displacement. Oscillations of complex systems, if excited with sufficient gentleness, can be approximated by small oscillations. This approximation leads to linearization, which allows the decomposition of complex oscillations of a system into independent normal modes of fixed frequencies. In this chapter, we will learn about small oscillations, linearization, and normal modes.

2.2 Simple pendulum and linearization

To introduce the concept of linearization, we consider the well-known example of a simple pendulum; i.e., a point mass m hanging by a flexible and massless string of length ℓ.

Example 2.1 Simple pendulum

Derive the Lagrange's equation of motion for the simple pendulum shown in Fig. 2.1, and solve it under small angular amplitude approximation.

The Lagrangian for this system is given by

$$L = \frac{1}{2} m\ell^2 \dot{\theta}^2 - m\, g\, \ell\, (1 - \cos\theta) \,, \tag{2.1}$$

where θ denotes the angle made by the string with the vertical. The potential energy is chosen to be zero at the equilibrium point of the bob. Since the amplitude θ is assumed to be small, we approximate $\cos\theta$ by $1 - \theta^2/2$ and drop all higher order terms, which ensures that the generalized force $-\frac{\partial V}{\partial \theta}$ is

DOI: 10.1201/9781003174967-2

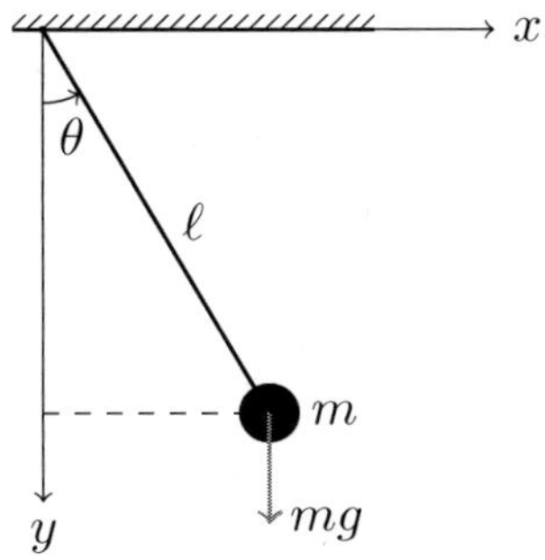

FIGURE 2.1
A point mass m hanging by a flexible massless string of length ℓ.

linear in θ. This linearization leads to

$$L = \frac{1}{2}\, m\ell^2\, \dot{\theta}^2 - \frac{1}{2}\, m\, g\, \ell\, \theta^2 \; , \tag{2.2}$$

and consequently the equation of motion for θ is then given by

$$\ddot{\theta} + \frac{g}{\ell}\, \theta = 0 \; . \tag{2.3}$$

The general solution of Eq. (2.2) is $\theta(t) = \theta_0 \sin\left(\sqrt{g/\ell}\; t + \delta\right)$. The angle θ_0 is the angular amplitude and δ denotes the initial phase. Note that the partial derivative with respect to $\dot{\theta}$ was carried out keeping θ constant, and *vice versa*[1].

2.3 Linearization: the general case

Let us consider a system with n-degrees of freedom, described by a set of generalized coordinates $\{\tilde{q}_1, \tilde{q}_2, \cdots \tilde{q}_n\} \equiv \tilde{q}$, which is in oscillation about a stable equilibrium configuration described by $q_0 \equiv \{q_{0_1}, q_{0_2}, \cdots q_{0_n}\}$. Coordinates $q_i = \tilde{q}_i - q_{0_i}$ measure the displacements from equilibrium configuration. We will now write down the general form of kinetic and potential energies for this system.

[1]See Ref. [15, 29] for examples of absurdities we may be led to if we are not careful about these partial derivatives.

2.3.1 Kinetic and potential energies

As we saw in Sec. (1.4), in terms of the generalized coordinates q_i and the corresponding velocities $\dot{q}_i$, when the coordinates q_i do not depend explicitly on time, the kinetic energy can be written as

$$T = \frac{1}{2}\sum_{ij} T_{ij}(q)\,\dot{q}_i\dot{q}_j \ , \qquad i,j = 1,\ldots,n, \tag{2.4}$$

where $i,j = 1,2,\ldots,n$, and n is the number of degrees of freedom. Thus, the kinetic energy T can be assumed to be a homogeneous quadratic function of generalized velocities $\dot{q}_i$. The coefficients $T_{ij}(q)$ are, in general, functions of the coordinates. For small oscillations we replace them with their values at the equilibrium point; i.e., $T = \frac{1}{2}\sum_{ij} T_{ij}(q)\,\dot{q}_i\dot{q}_j \approx \frac{1}{2}\sum_{ij} T_{ij}\,\dot{q}_i\dot{q}_j$, where $T_{ij} = T_{ij}(q_0)$.

The potential energy $V(q)$ of the system near the equilibrium point q_0 can be written as

$$V(q) = V(q_o) + \sum_i \left.\frac{\partial V}{\partial q_i}\right|_{q_0} q_i + \frac{1}{2}\sum_{ij} \left.\frac{\partial^2 V}{\partial q_j \partial q_i}\right|_{q_0} q_i\, q_j + \mathcal{O}(|q|^3) \tag{2.5}$$

where all derivatives are evaluated at the equilibrium point. Since the forces are zero at the equilibrium point, all first derivative terms vanish. Hence, choosing $V(q_0) = 0$, the potential energy becomes

$$V(q) = \frac{1}{2}\sum_{ij} \left.\frac{\partial^2 V}{\partial q_j \partial q_i}\right|_{q_0} q_i\, q_j + \mathcal{O}(|q|^3)\ . \tag{2.6}$$

We restrict ourselves to "small oscillations"; i.e., we will assume that terms that are $\mathcal{O}(|q|^3)$ or higher are negligible with respect to the second order terms. Thus the small-oscillation approximation reduces the potential energy of the system to a *homogeneous quadratic function* of coordinates:

$$V(q) \approx \frac{1}{2}\sum_{ij} \left.\frac{\partial^2 V}{\partial q_j \partial q_i}\right|_{q_0} q_i\, q_j \equiv \frac{1}{2}\sum_{ij} V_{ij}\, q_i\, q_j\ .$$

This reduction of the potential to a quadratic homogeneous function of the coordinates is known as the linearization of the system, as it results in equations of motion that are linear in coordinates and their derivatives.

2.3.2 Kinetic and potential energies in matrix notation

We define a $n \times n$ constant matrix $\mathbb{T}$ whose ij-th element is T_{ij}. Then we can write the kinetic energy of the system as

$$
\begin{aligned}
T &= \frac{1}{2}\sum_{ij} T_{ij}\,\dot{q}_i\dot{q}_j = \frac{1}{2}\,(\dot{q}_1, \cdots, \dot{q}_n)\begin{pmatrix} T_{11} & \cdots & T_{1n} \\ \vdots & \cdots & \vdots \\ T_{n1} & \cdots & T_{nn}\end{pmatrix}\begin{pmatrix}\dot{q}_1 \\ \vdots \\ \dot{q}_n\end{pmatrix} \\
&= \frac{1}{2}\,\dot{q}^{\mathrm{T}}\mathbb{T}\dot{q}\ , \qquad (2.7)
\end{aligned}
$$

where $\dot{q}$ is the time derivative of the $n \times 1$ column vector

$$
q \equiv \begin{pmatrix} q_1 \\ \vdots \\ q_n \end{pmatrix},
$$

and $\dot{q}^T$ represents its transpose.

Example 2.2 The $\mathbb{T}$ matrix for a free particle
For a free particle, describe the $\mathbb{T}$-matrix in Cartesian and spherical coordinates.

- **Cartesian coordinates:** In Cartesian coordinates, the kinetic energy is

$$
T = \frac{1}{2}mv^2 = \frac{1}{2}m(\dot{x}^2 + \dot{y}^2 + \dot{z}^2)\ .
$$

The corresponding $\mathbb{T}$ matrix is

$$
\mathbb{T} = \begin{pmatrix} m & 0 & 0 \\ 0 & m & 0 \\ 0 & 0 & m \end{pmatrix},
$$

which is proportional to the unit matrix; i.e., $\mathbb{T} = m \times \mathbb{1}$.

- **Spherical coordinates:** As shown in Fig. 2.2, the Cartesian coordinates and the spherical coordinates are related by

$$
\begin{aligned}
x &= r \sin\theta \cos\varphi\ , \\
y &= r \sin\theta \sin\varphi\ , \qquad (2.8) \\
z &= r \cos\theta\ .
\end{aligned}
$$

The kinetic energy is then given by

$$
\begin{aligned}
T &= \frac{1}{2}mv^2 = \frac{1}{2}m(\dot{x}^2 + \dot{y}^2 + \dot{z}^2) \\
&= \frac{1}{2}\left(m\dot{r}^2 + mr^2\dot{\theta}^2 + mr^2\sin^2\theta\,\dot{\varphi}^2\right)\ , \qquad (2.9)
\end{aligned}
$$

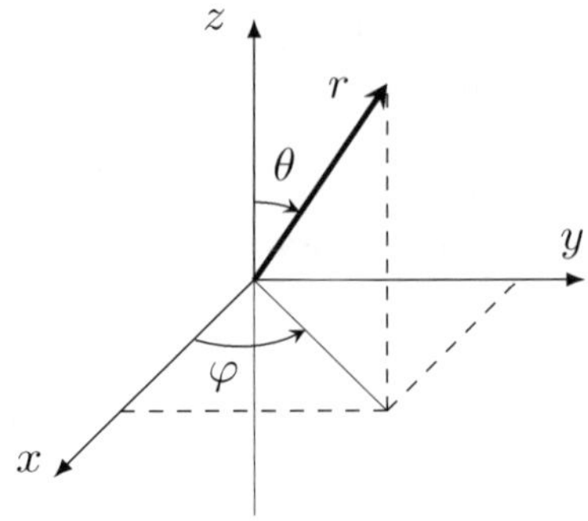

FIGURE 2.2
The position of a particle described by spherical coordinates.

The matrix $\mathbb{T}$ is again diagonal due to the orthogonal nature of these coordinates. However, it is no longer proportional to the unit matrix and its entries are not constant. The $\mathbb{T}$ matrix is given by

$$\mathbb{T} = \begin{pmatrix} m & 0 & 0 \\ 0 & mr^2 & 0 \\ 0 & 0 & mr^2 \sin^2\theta \end{pmatrix} .$$

Since, for a small-oscillation problem, the values of the coordinates are not expected to change much from their values at the equilibrium, the matrix elements are evaluated at q_0: $T_{ij}(q) = T_{ij}(q_0)$. Thus, for the case above, the T_{ij}'s are no longer functions of coordinates and are instead given by

$$\mathbb{T} = \begin{pmatrix} m & 0 & 0 \\ 0 & mr_0^2 & 0 \\ 0 & 0 & mr_0^2 \sin^2\theta_0 \end{pmatrix} .$$

Similar to $\mathbb{T}$, we define a matrix $\mathbb{V}$ whose ij-th element is V_{ij}, and hence the potential energy of the system can be written as

$$\begin{aligned} V &= \frac{1}{2}\sum_{ij} V_{ij}\, q_i q_j = \frac{1}{2}\,(q_1, \cdots, q_n) \begin{pmatrix} V_{11} & \cdots & V_{1n} \\ \vdots & \cdots & \vdots \\ V_{n1} & \cdots & V_{nn} \end{pmatrix} \begin{pmatrix} q_1 \\ \vdots \\ q_n \end{pmatrix} \\ &= \frac{1}{2} q^{\mathrm{T}} \mathbb{V} q \, . \end{aligned} \tag{2.10}$$

We next determine the $\mathbb{T}$ and $\mathbb{V}$ matrices for a several examples.

Example 2.3 The $\mathbb{T}$ and $\mathbb{V}$ matrices for a diatomic molecule
Determine $\mathbb{T}$ and $\mathbb{V}$ matrices for a system comprising two identical particles of mass m and connected by a spring of constant k, as shown in Fig. 2.3.

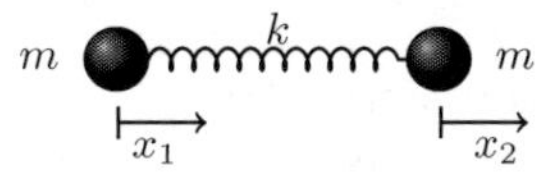

FIGURE 2.3
Two identical particles of mass m connected by a spring of stiffness k.

We denote the positions of the two balls from their equilibrium points by coordinates x_1 and x_2. Then, the kinetic and potential energies are given by

$$T = \frac{1}{2}m\left(\dot{x}_1^2 + \dot{x}_2^2\right) \quad \text{and} \quad V = \frac{1}{2}k\left(x_2 - x_1\right)^2 . \tag{2.11}$$

Hence the corresponding matrices are

$$\mathbb{T} = \begin{pmatrix} m & 0 \\ 0 & m \end{pmatrix} \quad \text{and} \quad \mathbb{V} = \begin{pmatrix} k & -k \\ -k & k \end{pmatrix} . \tag{2.12}$$

We now consider another system with two degrees of freedom, but this time one in which $\mathbb{T}$ has non-zero off-diagonal elements.

Example 2.4 Double pendulum: a case of off-diagonal $\mathbb{T}$ matrix
Determine $\mathbb{T}$ and $\mathbb{V}$ matrices for the double pendulum discussed in Example 1.3 and illustrated in Fig. 2.4.

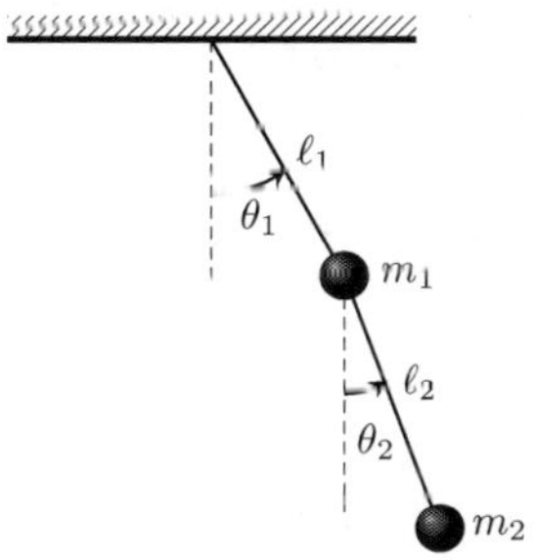

FIGURE 2.4
Double pendulum.

The kinetic energy for the system is given by

$$T = \frac{1}{2}(m_1 + m_2)\,\ell_1^2\,\dot{\theta}_1^{\,2} + \frac{1}{2}m_2\,\ell_2^2\,\dot{\theta}_2^{\,2} + \dot{\theta}_1\dot{\theta}_2\,m_2\,\ell_1\ell_2\cos(\theta_1 - \theta_2) .$$

Since in the equilibrium position both strings are vertical, we evaluate the coefficients T_{ij} at $\theta_1 = \theta_2 = 0$. This yields

$$T = \frac{1}{2}(m_1 + m_2)\,\ell_1^2\,\dot{\theta_1}^2 + \frac{1}{2}m_2\,\ell_2^2\,\dot{\theta_2}^2 + \frac{1}{2}\left(\dot{\theta}_1\dot{\theta}_2 + \dot{\theta}_2\dot{\theta}_1\right)\,m_2\,\ell_1\ell_2 \quad (2.13)$$

Note that we have deliberately symmetrized the product of $\dot{\theta_1}$ and $\dot{\theta_2}$ in the last term, a procedure that can be carried out more generally for most systems. Now, a comparison with Eq. (2.7) yields

$$\mathbb{T} = \begin{pmatrix} (m_1 + m_2)\ell_1^2 & m_2\ell_1\ell_2 \\ m_2\ell_1\ell_2 & m_2\ell_2^2 \end{pmatrix} \quad (2.14)$$

The potential energy of the system is

$$V(\theta_1, \theta_2) = -(m_1 + m_2)\,g\ell_1\cos\theta_1 - m_2\,g\,\ell_2\cos\theta_2.$$

For very small values of θ_1 and θ_2, expanding this potential we get

$$\begin{aligned} V(\theta_1, \theta_2) &\approx -(m_1 + m_2)g\ell_1\left(1 - \frac{\theta_1^2}{2}\right) - m_2 g\ell_2\left(1 - \frac{\theta_2^2}{2}\right) \\ &\approx -(m_1 + m_2)g\ell_1 - m_2 g\ell_2 + \frac{1}{2}(m_1 + m_2)g\ell_1\theta_1^2 + \frac{1}{2}m_2 g\ell_2\theta_2^2 \\ &\approx V_0 + \frac{1}{2}V_{ij}\theta_i\theta_j \end{aligned} \quad (2.15)$$

where $V_0 = -(m_1 + m_2)g\ell_1 - m_2 g\ell_2$, and V_{ij} are the elements the matrix $\mathbb{V}$

$$\mathbb{V} = \begin{pmatrix} (m_1 + m_2)g\ell_1 & 0 \\ 0 & m_2 g\ell_2 \end{pmatrix}. \quad (2.16)$$

2.3.2.1 Symmetry properties of $\mathbb{V}$ and $\mathbb{T}$ matrices

Since the coefficients V_{ij} represent second order derivatives of a well-defined potential function, they are naturally symmetric in indices i and j, i.e., $V_{ij} = V_{ji}$, and hence $\mathbb{V} = \mathbb{V}^T$. We now show that the coefficients T_{ij} of kinetic energy can be assumed to be symmetric, in general.

An arbitrary matrix $\mathbb{T}$ can be written as a sum of a symmetric matrix $\mathbb{T}^S$ and an antisymmetric matrix $\mathbb{T}^A$ as follows:

$$\begin{aligned} \mathbb{T}_{ij} &= \frac{1}{2}\left(\mathbb{T}_{ij} + \mathbb{T}_{ji}\right) + \frac{1}{2}\left(\mathbb{T}_{ij} - \mathbb{T}_{ji}\right) \\ &= \mathbb{T}_{ij}^S + \mathbb{T}_{ij}^A\,, \end{aligned} \quad (2.17)$$

where $\mathbb{T}^S_{ij} = \frac{1}{2}(\mathbb{T}_{ij} + \mathbb{T}_{ji})$ and $\mathbb{T}^A_{ij} = \frac{1}{2}(\mathbb{T}_{ij} - \mathbb{T}_{ji})$ are coefficients of symmetric and antisymmetric matrices respectively. We will show that $\mathbb{T}^A$, the antisymmetric part of $\mathbb{T}$, does not contribute to kinetic energy, and thus, without loss of generality, $\mathbb{T}$ can always be assumed to be a symmetric matrix.

The contribution of the antisymmetric part to the kinetic energy $\sum_{ij} \mathbb{T}^A_{ij}\dot{q}^i\dot{q}^j$ can be written as:

$$\sum_{ij} \mathbb{T}^A_{ij}\dot{q}_i\,\dot{q}_j = \frac{1}{2}\sum_{ij} \mathbb{T}^A_{ij}\,\dot{q}_i\,\dot{q}_j + \frac{1}{2}\sum_{ij} \mathbb{T}^A_{ij}\,\dot{q}_i\,\dot{q}_j\ .$$

Since i and j are arbitrary indices or *dummy indices*, we change the order of the indices for the second term and get

$$\sum_{ij} \mathbb{T}^A_{ij}\,\dot{q}_i\,\dot{q}_j = \frac{1}{2}\sum_{ij} \mathbb{T}^A_{ij}\,\dot{q}_i\,\dot{q}_j + \frac{1}{2}\sum_{ij} \mathbb{T}^A_{ji}\,\dot{q}_j\,\dot{q}_i = \frac{1}{2}\sum_{ij} \dot{q}_i\,\dot{q}_j\,(\mathbb{T}^A_{ij} + \mathbb{T}^A_{ji}) = 0\ .$$

Hence, the decomposition of the kinetic energy into symmetric and antisymmetric parts leads to

$$\sum_{ij} \mathbb{T}_{ij}\,\dot{q}_i\,\dot{q}_j \quad = \quad \sum_{ij} \mathbb{T}^S_{ij}\,\dot{q}_i\,\dot{q}_j + \sum_{ij} \mathbb{T}^A_{ij}\,\dot{q}_j\,\dot{q}_i = \sum_{ij} \mathbb{T}^S_{ij}\dot{q}_i\,\dot{q}_j\ . \tag{2.18}$$

Thus, without loss of generality, we can choose $\mathbb{T}$ to be symmetric, i.e., $\mathbb{T} = \mathbb{T}^T$.

2.4 Normal coordinates and decoupling of modes

We will now revisit Example (2.3) and derive its equations of motion, which turn out to be two coupled differential equations. We then decouple these differential equations to determine the state of the system as it evolves from its initial condition at time $t = 0$.

Example 2.5 The diatomic molecule revisited
For the system of Example (2.3), determine the equations of motion and decouple them. Assume that at $t = 0$, one of the masses is hit by a hammer and the resulting motion is characterized by the following set of initial conditions: $x_1 = x_2 = 0$ and $\dot{x}_1 = v_0 > 0$ and $\dot{x}_2 = 0$.

The Lagrangian of this system is

$$L = \frac{1}{2}m\left(\dot{x}_1^2 + \dot{x}_2^2\right) - \frac{1}{2}k\,(x_2 - x_1)^2\ . \tag{2.19}$$

The equations of motion $\frac{d}{dt}\frac{\partial L}{\partial \dot{x}_i} - \frac{\partial L}{\partial x_i} = 0$ with $i = 1, 2$, then generate the following two coupled differential equations:

$$m\ddot{x}_1 + k\,(x_1 - x_2) = 0 \qquad \text{and} \qquad m\ddot{x}_2 + k\,(x_2 - x_1) = 0\ . \tag{2.20}$$

To decouple these equations, we introduce two new variables

$$\xi_+ = \frac{1}{2}(x_1 + x_2) \qquad \text{and} \qquad \xi_- = \frac{1}{2}(x_1 - x_2)\ . \tag{2.21}$$

The sum and the difference of Eqs. (2.20) can then be expressed in terms $\xi_\pm$ as

$$m\,\ddot{\xi}_+ = 0 \qquad \text{and} \qquad m\,\ddot{\xi}_- + 2k\,\xi_- = 0. \tag{2.22}$$

These judiciously chosen variables that decouple the equations of motion are known as the *normal coordinates* of the system. The general solutions for $\xi_\pm$ are: $\xi_+ = A\,t + B$ and $\xi_- = R\cos(\omega\,t + \delta)$, where $\omega = \sqrt{2k/m}$. Constants A, B, R, and δ will now be determined from the initial positions and velocities of the two particles. Substituting the above initial conditions, we get, $A = \frac{1}{2}v_0$, $B = 0$, $R = \frac{1}{2}v_0/\omega$ and $\delta = -\pi/2$. Hence,

$$\xi_+ = \frac{1}{2}v_0\,t \qquad \text{and} \qquad \xi_- = \frac{v_0}{2\omega}\sin\omega\,t\ , \tag{2.23}$$

and hence, from Eqs. (2.21)

$$x_1 = \frac{v_0}{2}\left(t + \frac{\sin\omega\,t}{\omega}\right) \quad \text{and} \quad x_2 = \frac{v_0}{2}\left(t - \frac{\sin\omega\,t}{\omega}\right)\ . \tag{2.24}$$

In this example, the normal coordinate ξ_+ denotes the position of the center of mass of the system that moves with constant velocity $v_0/2$. In contrast, the normal coordinate ξ_- captures the interesting aspect of the motion which is relative oscillation of the two masses.

Exercise 2.1 For Example (2.5), show that at $t = 0$, Eq. (2.24) generates the initial conditions given in the example.

This example had sufficient symmetry that allowed us to guess the two normal coordinates ξ_+ and ξ_-, decouple the differential equations, and solve them independently. For more complex systems, it would be difficult to identify the normal coordinates that would decouple the equations of motion. We now develop a general formalism for determining the normal coordinates for small oscillations of complex systems.

2.4.1 Euler-Lagrange's equations for small oscillations

For a system subjected to small oscillations, the Lagrangian is given by

$$L = T - V = \frac{1}{2}\sum_{ij} T_{ij}\,\dot{q}_i\dot{q}_j \;-\; \frac{1}{2}\sum_{ij} V_{ij}\,q_iq_j\ , \tag{2.25}$$

where T_{ij} and V_{ij} are constants. The Euler-Lagrange's equations are then given by

$$\sum_j (T_{ij}\,\ddot{q}_j \; + \; V_{ij}q_j) = 0 \;, \qquad i = 1, 2, \ldots, n \;, \tag{2.26}$$

which could be succinctly written in matrix notation as

$$\mathbb{T}\ddot{q} \; + \; \mathbb{V}q = 0 \;. \tag{2.27}$$

In general, $\mathbb{T}$ and $\mathbb{V}$ are real symmetric matrices with non-zero off-diagonal terms, and hence Eqs. (2.27) represent n coupled linear differential equations for coordinates q_i and their time-derivatives $\ddot{q}_i$. Our objective is to decouple these differential equations. For this, we seek new coordinates ξ_k such that the $\mathbb{T}$ and $\mathbb{V}$ matrices become diagonal when q_i are expressed as linear combinations of ξ_k. Coordinates ξ_k are called *normal coordinates* .

Since the real and symmetric nature of $\mathbb{T}$ and $\mathbb{V}$ matrices play very important roles in the decoupling process, we briefly review their mathematical properties that are needed for this section.

2.4.2 On the eigenvalues of the $\mathbb{T}$ and $\mathbb{V}$ matrices

We first discuss the eigenvalues of the $\mathbb{T}$ matrix. Since $\mathbb{T}$ is a real and symmetric matrix, it can be diagonalized by an orthogonal matrix $\mathbb{O}$; i.e., $\mathbb{O}\mathbb{T}\mathbb{O}^T = \mathbb{D}$, where $\mathbb{D}$ is a diagonal matrix with $\mathbb{D}_{k\ell} = \lambda_k \delta_{k\ell}$, and λ_k are the eigenvalues of $\mathbb{T}$. We write then the kinetic energy as

$$\begin{aligned} T &= \frac{1}{2}\sum_{ij} T_{ij}\,\dot{q}_i\dot{q}_j = \frac{1}{2}\,\dot{q}^{\mathrm{T}}\mathbb{T}\dot{q} = \frac{1}{2}\,\left(\dot{q}^{\mathrm{T}}\mathbb{O}^T\right)\left(\mathbb{O}\mathbb{T}\mathbb{O}^T\right)\left(\mathbb{O}\dot{q}\right) \\ &= \frac{1}{2}\,\dot{\eta}^{\mathrm{T}}\mathbb{D}\dot{\eta} = \frac{1}{2}\sum_k \lambda_k\,(\dot{\eta}_k)^2 \;, \end{aligned}$$

where $\dot{\eta} = \mathbb{O}\dot{q}$. Note that λ_k cannot be equal to zero, as otherwise the diagonalization would result in the description of the system by a smaller number of coordinates than we began with, which is the number of degrees of freedom of the system. Furthermore, none of the eigenvalues could be negative either, as kinetic energy is always a positive quantity.

For the eigenvalues of the $\mathbb{V}$ matrix, we note that they also cannot be negative. The possibility of a zero-eigenvalue for $\mathbb{V}$ allows for the uniform motion of the center of mass, as we saw in Example (2.5).

2.4.3 Decoupling of equations of motion – the general case

In Example (2.3) we decoupled the differential equations for x_1 and x_2 by judiciously choosing new generalized coordinates $\xi_\pm = (x_1 \pm x_2)/2$. These special combinations $\xi_\pm$ were examples of the normal coordinates for that system. In this section we will learn the general procedure for finding these normal coordinates that would decouple the following differential equations:

$$\mathbb{T}\ddot{q} + \mathbb{V}q = 0 . \tag{2.28}$$

To solve these coupled differential equations, we first seek oscillatory solutions of Eq. (2.28) in the form of

$$\eta = \eta(0) \cos(\omega t + \delta) , \tag{2.29}$$

where $\eta(0)$ represents a $n \times 1$ column vector [2]. Substituting in Eq. (2.28) , we get

$$\mathbb{V}\eta = \omega^2 \mathbb{T}\eta , \qquad \text{or} \qquad \left[\mathbb{V} - \omega^2 \mathbb{T}\right] \eta = 0 . \tag{2.30}$$

Eq. (2.30) represents n homogeneous linear equations that allow us to determine the n components of the vector η. A non-zero solution exists only if

$$\text{Det} \left[\mathbb{V} - \omega^2 \, \mathbb{T}\right] = 0 . \tag{2.31}$$

Solving Eq. (2.31), we obtain n normal frequencies ω_i^2. The corresponding normal coordinate vectors η_i, which are $n \times 1$ column vectors, are then determined by

$$\left[\mathbb{V} - \omega_i^2 \, \mathbb{T}\right] \eta_i = 0 , \quad \eta_i \equiv \begin{bmatrix} \eta_{i1} \\ \eta_{i2} \\ \vdots \\ \eta_{in} \end{bmatrix} . \tag{2.32}$$

All of the frequencies ω_i^2 are real and positive (or zero) and the corresponding normal coordinate vectors satisfy the following orthonormality condition

$$\left(\eta_i\right)^T \mathbb{T} \eta_j = \delta_{ij} . \tag{2.33}$$

For additional details, see Appendix II, Sec. (2.7).

We will now use these normal coordinate vectors to determine the normal coordinates ξ_k, as linear combinations of q_i, that decouple the equations of motion. We first note that

$$q \equiv \begin{bmatrix} q_1 \\ q_2 \\ \vdots \\ q_n \end{bmatrix} = \sum_k \begin{bmatrix} \eta_{k1} \\ \eta_{k2} \\ \vdots \\ \eta_{kn} \end{bmatrix} A_k \cos(\omega_k t + \delta_k) \equiv \sum_k \eta_k \, \xi_k(t) . \tag{2.34}$$

[2]See Appendix II (2.7), for a proof of existence of such common eigenvectors.

We next multiply Eq. (2.34) from left by matrix $\mathbb{T}$, which gives

$$\mathbb{T}\, q = \sum_k \mathbb{T}\, \eta_k\, \xi_k(t)\ .$$

Then, Eq. (2.33) leads to

$$\xi_k = \eta_k^T\, \mathbb{T}\, q\ . \tag{2.35}$$

We will now see several examples that elaborate upon the above decoupling formalism.

Example 2.6 A triatomic molecule
Let us consider a system with three equal masses m that are connected by two identical springs with spring constant k, as shown in Fig. 2.5. The positions of these masses, measured from their respective equilibrium points, are denoted by x_1, x_2 and x_3.

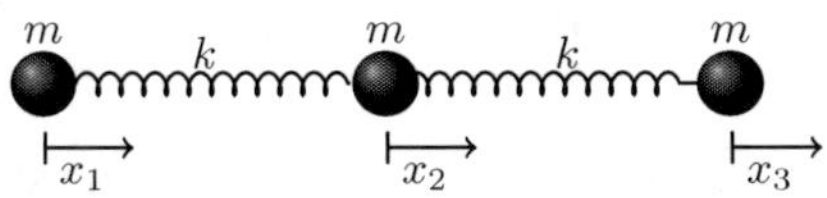

FIGURE 2.5
Three equal masses m connected by two identical springs of constant k.

The kinetic energy and the potential energy of the system are given by

$$\begin{aligned} T &= \frac{1}{2}\, m(\dot{x}_1^2 + \dot{x}_2^2 + \dot{x}_3^2) \quad \text{and} \\ V &= \frac{1}{2}\, k(x_1 - x_2)^2 + \frac{1}{2}\, k(x_2 - x_3)^2 \\ &= \frac{1}{2}\, k(x_1^2 + 2x_2^2 + x_3^2 - 2x_1x_2 - 2x_2x_3)\ . \end{aligned} \tag{2.36}$$

The $\mathbb{T}$ and $\mathbb{V}$ matrices are then given by

$$\mathbb{T} = \begin{pmatrix} m & 0 & 0 \\ 0 & m & 0 \\ 0 & 0 & m \end{pmatrix} \qquad \text{and} \qquad \mathbb{V} = \begin{pmatrix} k & -k & 0 \\ -k & 2k & -k \\ 0 & -k & k \end{pmatrix}\ . \tag{2.37}$$

The equations of motion in matrix form become

$$\mathbb{T}\ddot{x} + \mathbb{V}x = \begin{pmatrix} m & 0 & 0 \\ 0 & m & 0 \\ 0 & 0 & m \end{pmatrix} \begin{bmatrix} \ddot{x}_1 \\ \ddot{x}_2 \\ \ddot{x}_3 \end{bmatrix} + \begin{pmatrix} k & -k & 0 \\ -k & 2k & -k \\ 0 & -k & k \end{pmatrix} \begin{bmatrix} x_1 \\ x_2 \\ x_3 \end{bmatrix} = 0\ . \tag{2.38}$$

Since we seek oscillatory solutions, following Eq. (2.31), we set

$$\begin{aligned}\text{Det}\left[\mathbb{V}-\omega^2\,\mathbb{T}\right] &= \text{Det}\begin{bmatrix} k-m\omega^2 & -k & 0 \\ -k & 2k-m\omega^2 & -k \\ 0 & -k & k-m\omega^2 \end{bmatrix} \\ &= m\omega^2\left(m\omega^2-3k\right)\left(k-m\omega^2\right)=0\ . \end{aligned} \tag{2.39}$$

Therefore, the normal frequencies ω_i^2 are given by

$$\omega_1^2 = 0;\quad \omega_2^2 = \frac{3k}{m};\quad \text{and}\quad \omega_3^2 = \frac{k}{m}\ . \tag{2.40}$$

The corresponding normal coordinate vectors are:

$$\eta_1 = \begin{bmatrix} 1 \\ 1 \\ 1 \end{bmatrix};\quad \eta_2 = \begin{bmatrix} 1 \\ -2 \\ 1 \end{bmatrix};\quad \text{and}\quad \eta_3 = \begin{bmatrix} -1 \\ 0 \\ 1 \end{bmatrix}\ . \tag{2.41}$$

(a) Normal mode 2 (b) Normal mode 3

FIGURE 2.6
Two of the three modes described by the normal coordinates η_i.

These normal vectors allow us to determine the normal coordinates that would decouple the equations of motion given in Eq. (2.38). The first normal coordinate η_1 is then given by

$$\xi_1 = x_1 + x_2 + x_3\ . \tag{2.42}$$

This coordinate, belonging to the frequency $\omega_1^2 = 0$, represents the motion of the center of mass of the three-particle molecule and is not of much significance. The internal dynamics of the molecule are described by the second normal coordinate η_2

$$\xi_2 = x_1 - 2\,x_2 + x_3 \qquad \text{with } \omega_2^2 = \tfrac{3k}{m}\ , \tag{2.43}$$

and the third normal coordinate η_3

$$\xi_3 = x_1 - x_3 \qquad \text{with } \omega_2^2 = \tfrac{k}{m}\ . \tag{2.44}$$

These last two modes are depicted in Fig. 2.6.

Example 2.7 Normal modes for a double pendulum
Determine the normal coordinates, and corresponding frequencies, for the double-pendulum shown below. The $\mathbb{T}$ and $\mathbb{V}$ matrices, calculated in Example (2.4), are:

$$\mathbb{T} = \begin{pmatrix} 2m\ell^2 & m\ell^2 \\ m\ell^2 & m\ell^2 \end{pmatrix} \quad \text{and} \quad \mathbb{V} = \begin{pmatrix} 2mg\ell & 0 \\ 0 & mg\ell \end{pmatrix} . \tag{2.45}$$

To determine the normal frequencies, we solve the equation

$$\begin{aligned} \text{Det}\left[\mathbb{V} - \omega^2\mathbb{T}\right] &= \text{Det}\begin{bmatrix} 2\omega^2 l^2 m - 2glm & \omega^2 l^2 m \\ \omega^2 l^2 m & \omega^2 l^2 m - glm \end{bmatrix} \\ &= 2g^2 l^2 m^2 - 4g\omega^2 l^3 m^2 + \omega^4 l^4 m^2 = 0 . \end{aligned} \tag{2.46}$$

The two solutions are

$$\omega_1^2 = \left(2 + \sqrt{2}\right)\frac{g}{l} \quad \text{and} \quad \omega_2^2 = \left(2 - \sqrt{2}\right)\frac{g}{l} . \tag{2.47}$$

The corresponding normal coordinate vectors are determined by solving $\left[\mathbb{V} - \omega_i^2\mathbb{T}\right]\eta_i = 0$ and are respectively given by

$$\eta_1 = B_1 \begin{bmatrix} 1 \\ -\sqrt{2} \end{bmatrix} \quad \text{and} \quad \eta_2 = B_2 \begin{bmatrix} 1 \\ \sqrt{2} \end{bmatrix} . \tag{2.48}$$

where $B_1 = \left[m\ell^2(4 - 2\sqrt{2})\right]^{-1/2}$ and $B_2 = \left[m\ell^2(4 + 2\sqrt{2})\right]^{-1/2}$ are coefficients chosen such that $(\eta_1)^T\,\mathbb{T}\,\eta_1 = (\eta_2)^T\,\mathbb{T}\,\eta_2 = 1$. Note that $(\eta_1)^T\,\mathbb{T}\,\eta_2 = 0$, and thus η_1 and η_2 satisfy Eq. (2.33).

The normal coordinates are then determined as follows:

$$\begin{aligned} \begin{bmatrix} \theta_1 \\ \theta_2 \end{bmatrix} &= \begin{bmatrix} 1 \\ -\sqrt{2} \end{bmatrix} \underbrace{A_1\, cos(\omega_1 t + \delta_1)}_{\xi_1(t)} + \begin{bmatrix} 1 \\ +\sqrt{2} \end{bmatrix} \underbrace{A_2\, cos(\omega_2 t + \delta_2)}_{\xi_2(t)} \\ &= \begin{bmatrix} \xi_1(t) + \xi_2(t) \\ \sqrt{2}(\xi_2(t) - \xi_1(t)) \end{bmatrix} \end{aligned} \tag{2.49}$$

Solving Eqs. (2.49) for ξ_i, we get

$$\xi_1 = C_1\left(\sqrt{2}\,\theta_1 - \theta_2\right) \quad \text{and} \quad \xi_2 = C_2\left(\sqrt{2}\,\theta_1 + \theta_2\right) . \tag{2.50}$$

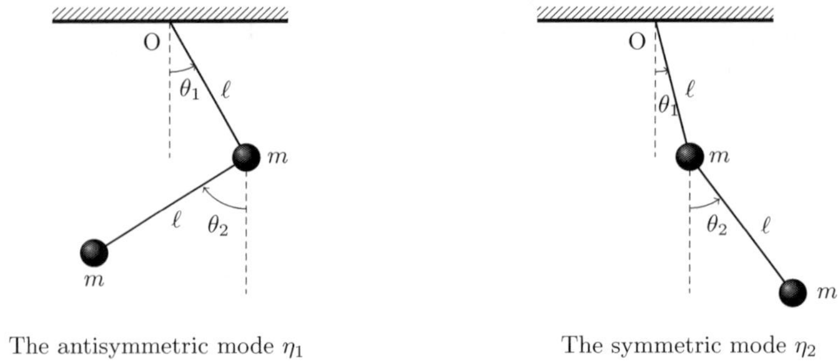

FIGURE 2.7
The antisymmetric and symmetric modes of a double pendulum.

For arbitrary non-zero values of coefficients C_i, these linear combinations of θ_1 and θ_2, will decouple Eq. (2.28) for this example. The two modes are depicted in Fig. 2.7.

Exercise 2.2 For Example (2.7), calculate the coefficients B_1 and B_2 such that vectors η_1 and η_2 obey $(\eta_1)^T\,\mathbb{T}\,\eta_1 = (\eta_2)^T\,\mathbb{T}\,\eta_2 = 1$

2.5 Problems

2.1 Determine the normal modes (frequencies) and corresponding eigenvectors for the two simple pendulums coupled a spring of constant k, as shown below.

2.2 For a particle of mass m, constrained to move on the surface of a paraboloid described by $x^2/a^2 + y^2/b^2 = z$, determine the $\mathbb{T}$ and $\mathbb{V}$ matrices for small oscillations near the origin. Assume gravitational acceleration to be $\vec{g} = -g\hat{z}$.

2.3 Determine the normal frequencies and corresponding normal vectors for the spring-pendulum system shown given in Fig. 2.9. If at time $t = 0$, all masses are rest, the spring is unstretched and ball is released from a small angle $\theta = \theta_0$, determine the displacement of the mass M and the angle θ at an arbitrary time $t > 0$. Investigate the motion of the system in the limit k tending to 0.

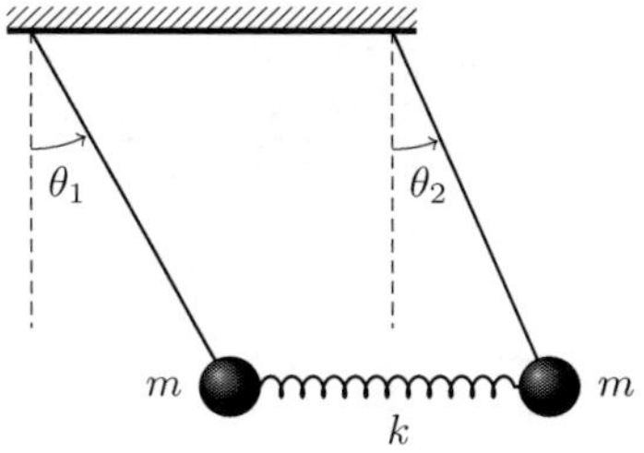

FIGURE 2.8
Figure for Problem 2.1: two simple pendulums of length ℓ coupled by a spring of constant k.

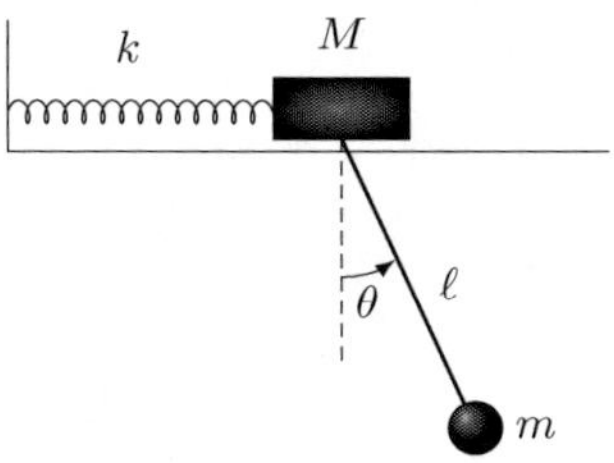

FIGURE 2.9
Figure for Problem 2.3: a spring-pendulum system.

2.4 Two balls of mass m are attached to a ball of mass M, and to walls, by four springs as shown in Fig. 2.10. Determine the normal frequencies and corresponding eigenvectors for the system. Coordinates x_i are measured from the equilibrium positions of the masses.

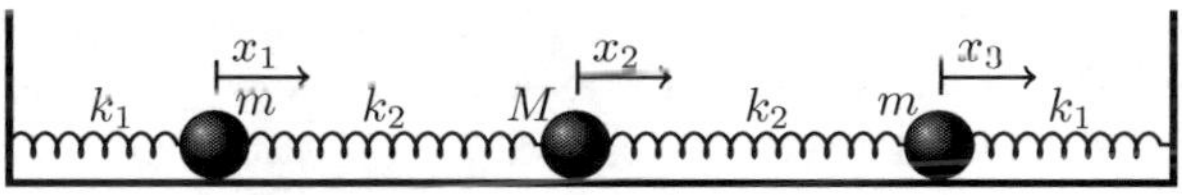

FIGURE 2.10
Three balls connected by springs.

2.5 Two identical objects of mass m, connected by a spring of constant k_2 are placed on a frictionless plane. One of the objects is attached to a wall with a spring of constant k_1 as shown in Fig. 2.11. Determine the normal modes of the system and corresponding frequencies. Coordinates x_1 and x_2 are measured from the equilibrium positions of the masses.

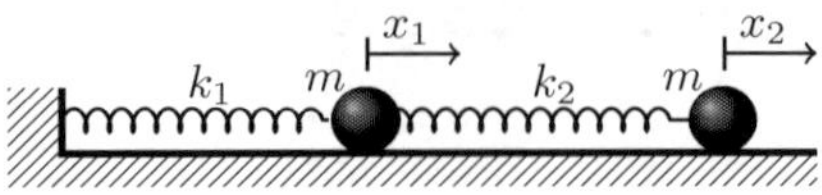

FIGURE 2.11
Two balls connected by springs.

2.6 Determine the normal frequencies and corresponding modes for the three identical beads of mass m, placed on a smooth (frictionless) ring of radius R, and connected to each other by identical springs of stiffness constant k, as shown below in Fig. 2.12. Denote the displacements of the beads from their equilibrium positions by angles θ_1, θ_2 and θ_3.

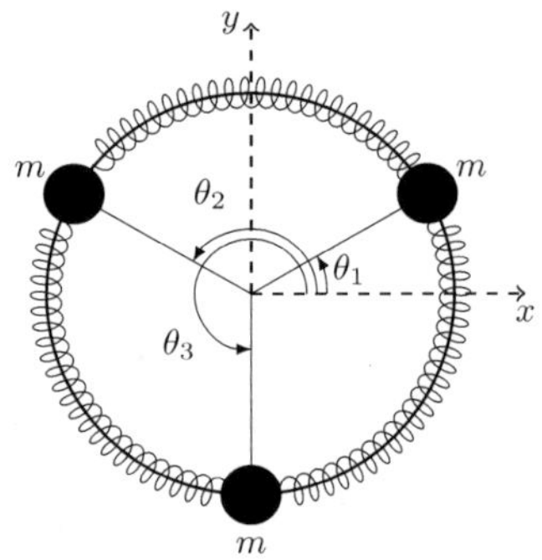

FIGURE 2.12
Figure for Problem 2.6: Beads on a frictionless ring connected by identical springs.

2.7 A physical pendulum, made of a uniform bar of mass m and length ℓ, is free to oscillate about a pivot at one end. The pivot point, tied to a spring of constant k, is constrained to move on a vertical axis and its position relative to the ceiling is denoted by y, as shown below. Determine the normal frequencies of oscillation and corresponding coordinates. Explore the motion in the limit $k \to \infty$.

2.8 A cylinder of radius R_1 is free to roll without slipping on the inner surface of a larger cylinder of radius R_2, as shown below. At equilibrium, both cylinders are horizontal and their axes lie on a vertical plane that bisects them both. The larger cylinder rolls on a horizontal surface without slipping. Use $\tilde{x}$ and θ to denote the generalized coordinate to describe the configuration of the system at any particular time. Determine normal modes of oscillations for the system.

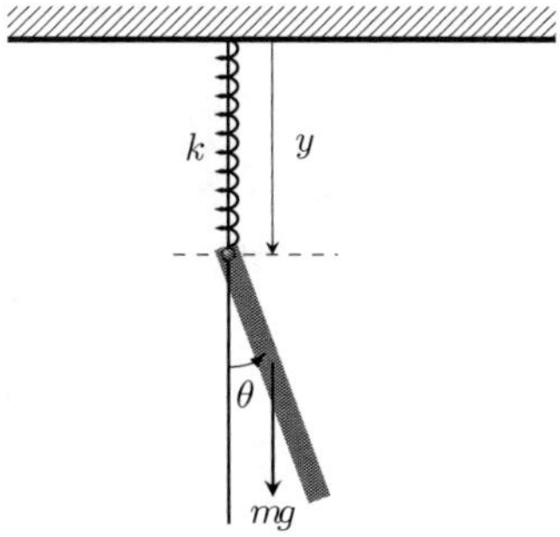

FIGURE 2.13
Figure for Problem 2.7: A physical pendulum tied to a vertical spring.

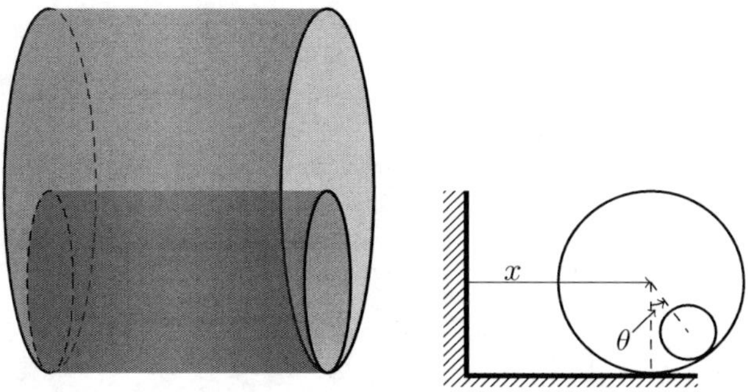

FIGURE 2.14
Figure for Problem 2.8.

2.6 Appendix I: properties of real symmetric matrices

We define a real symmetric matrix $\mathbb{S}$, a matrix whose elements obey

$$S_{ij} = S_{ji} \qquad \text{and} \qquad S^*_{ij} = S_{ij} \ , \tag{2.51}$$

where asterisk indicates complex conjugation. Here are some of the relevant properties of $\mathbb{S}$ which we list without proof.

1. A real symmetric matrix $\mathbb{S}$ can always be brought to a diagonal form by a similarity transformation generated by an appropriate orthogonal matrix $\mathbb{O}$, i.e.,

$$\mathbb{O}\,\mathbb{S}\,\mathbb{O}^T = \mathbb{D} \quad \text{and} \quad \mathbb{O}^T\mathbb{D}\,\mathbb{O} = \mathbb{S} \ , \tag{2.52}$$

where the matrix $\mathbb{O}$ satisfies $\mathbb{O}^T\mathbb{O} = \mathbb{O}\mathbb{O}^T = \mathbb{1}$.

2. Eigenvectors corresponding to distinct eigenvalues are orthogonal.

3. If Det[$\mathbb{S}$] $\neq 0$, then $\mathbb{S}^{-1}$ exists and is also a symmetric matrix.

4. The diagonal form $\mathbb{D}$ of $\mathbb{S}$ is given by

$$\mathbb{D} = \begin{pmatrix} \lambda_1 & 0 & \cdots & 0 \\ 0 & \lambda_2 & \cdots & 0 \\ & & \ddots & \\ 0 & 0 & \cdots & \lambda_n \end{pmatrix}$$

   where $\lambda_1, \cdots, \lambda_n$ are the eigenvalues of $\mathbb{S}$.

5. The determinants of $\mathbb{S}$ and $\mathbb{D}$ are equal and Det[$\mathbb{S}$] = Det[$\mathbb{D}$] $= \Pi_{i=1}^{n} \lambda_i$.

6. If all eigenvalues of the matrix $\mathbb{D}$ are positive, then a fractional power of $\mathbb{D}$ is defined by

$$\mathbb{D}^{\alpha} = \begin{pmatrix} \lambda_1^{\alpha} & 0 & \cdots & 0 \\ 0 & \lambda_2^{\alpha} & \cdots & 0 \\ & & \ddots & \\ 0 & 0 & \cdots & \lambda_n^{\alpha} \ . \end{pmatrix}$$

   In particular, the square root of $\mathbb{S}$ is then defined by

$$\mathbb{S}^{1/2} = \mathbb{O}^T \mathbb{D}^{1/2} \, \mathbb{O} \ .$$

   From this, it follows that $\mathbb{S}^{1/2}$, and its inverse $\mathbb{S}^{-1/2}$, are also real and symmetric matrices.

2.7 Appendix II: further remarks on the diagonalization process

Here we provide explicit steps to show how Eqs. (2.28) can be decoupled. Since we are seeking oscillating solutions, substituting $q = \eta \cos(\omega t + \delta)$, we get

$$\mathbb{V}\eta = \omega^2 \mathbb{T}\eta \ . \tag{2.53}$$

Due to the presence of $\mathbb{T}$ on the right, which is not necessarily a multiple of the unit matrix, Eq. (2.53) is not a typical eigenvalue problem. The properties of $\mathbb{T}$ and $\mathbb{V}$ matrices imply that, for any set of n-variables ξ_i, $\sum_{ij} T_{ij}\, \xi_i \xi_j = \xi^{\mathrm{T}} \mathbb{T} \xi > 0$, and $\sum_{ij} V_{ij}\, \xi_i \xi_j = \xi^{\mathrm{T}} \mathbb{V} \xi \geq 0$. We note that $\omega^2 = \frac{\eta^T \, \mathbb{V} \, \eta}{\eta^T \, \mathbb{T} \, \eta} \geq 0$, and hence all frequencies are positive or zero. However, as discussed in Sec. (2.4.2), the eigenvalues of $\mathbb{T}$ are all strictly positive, and this allows us to define both $\mathbb{T}^{\pm 1/2}$, which are real and symmetric matrices as well.

We first multiply Eq. (2.53) from the left by $\mathbb{T}^{-1/2}$ to get

$$\left(\mathbb{T}^{-1/2}\,\mathbb{V}\mathbb{T}^{-1/2}\right)\mathbb{T}^{1/2}\eta = \omega^2\mathbb{T}^{1/2}\eta. \tag{2.54}$$

We now define an intermediate set of n-coordinates $\chi = \mathbb{T}^{1/2}\eta$, which reduces the above equation to

$$\left(\mathbb{T}^{-1/2}\,\mathbb{V}\mathbb{T}^{-1/2}\right)\chi = \omega^2\chi. \tag{2.55}$$

Since $\mathbb{T}^{\pm 1/2}$ are real and symmetric matrices, the matrix $\mathbb{T}^{-1/2}\,\mathbb{V}\mathbb{T}^{-1/2}$ is also real and symmetric, and hence can be diagonalized by an orthogonal matrix $\mathbb{O}$. The resulting eigenvectors χ are then orthonormal, i.e., $\chi_i^T\chi_i = \delta_{ij}$. This implies the following orthonormality condition, weighted by matrix $\mathbb{T}$, that was stated in Eq. (2.33):

$$\left(\eta_i^T\mathbb{T}^{1/2}\right)\left(\mathbb{T}^{1/2}\eta_j\right) = \eta_i^T\mathbb{T}\eta_j = \delta_{ij}\ . \tag{2.56}$$

Part II

Electricity and Magnetism

3

Electricity and Magnetism: A Brief Review

3.1 Introduction

Magnetism, in form of *loadstones*, has been known for thousands of years. Electricity was also known in the ancient world. In particular, Amber (elektron in Greek) was known for its ability to pick up small objects when rubbed with cat's fur. However, the detailed quantitative investigation of these phenomena is relatively new. Nowadays, we know that these electric and magnetic properties are accurately described by corresponding fields that are created by electrical charge and current distributions respectively. Relationship among these fields and sources (charges and currents) are governed by Maxwell's equations – topic of this and the next few chapters. When the source distributions and boundary conditions are constant in time, the two fields would remain independent. In general, these fields are intricately related to each other. The richness of their inter-relationship becomes evident only when these fields are changing in time.

In this chapter, we briefly review concepts in electrostatics and magnetostatics, that would have been covered in detail in the first course in electricity and magnetism.

3.2 Review of electrostatics and magnetostatics

3.2.1 Coulomb's law

The force on an electric charge q subjected to an electric field $\vec{E}$ is given by

$$\vec{F} = q\vec{E} \ . \tag{3.1}$$

In fact, the above equation is used to define electric field at a point in space. I.e.,

$$\vec{E} = \lim_{q_o \to 0} \frac{\vec{F}}{q_o} \ , \tag{3.2}$$

where the limit indicates that the test charge q_o should be chosen as small as possible in order not to disturb the original field that it was supposed to

DOI: 10.1201/9781003174967-3

probe. In this section, we will look at fields created by various forms of charge distributions.

3.2.1.1 Coulomb's law: point charge

For a point charge, Coulomb's law states that electric field $\vec{E}$ at point $\vec{r}$ (*the field point*) due to a charge q at $\vec{r}_o$ or (*the source point*) is given by

$$\vec{E}(\vec{r}) = \frac{q}{4\pi\epsilon_o} \frac{\vec{r} - \vec{r}_o}{|\vec{r} - \vec{r}_o|^3} \, . \tag{3.3}$$

Thus, for a positive charge q, Eq. (3.3) implies that the $\vec{E}$-field at point $\vec{r}$, is directed radially away from the point $\vec{r}_o$, and for a negative charge it points towards $\vec{r}_o$. The constant ϵ_o is known as the *permittivity of free space*, and its numerical value depends upon the units used to measure electric field, charge and distance. In SI units, its value is $\epsilon_o = 8.8542 \times 10^{-12}\,\mathrm{C^2\,m^{-2}\,N^{-1}}$.

Exercise 3.1 Determine the force on a $2\mu C$ charge situated at (3,4) due to a $-3\mu C$ charge at (7,1), where coordinate points indicate positions relative to the origin, measured in meters.

3.2.1.2 Coulomb's law: continuous charge density

For a continuous charge density ρ distributed over a volume $\mathcal{V}$, replacing q by $\rho(\vec{r}\,')\,d^3r'$, we get

$$\vec{E}(\vec{r}) = \frac{1}{4\pi\epsilon_o} \int_{\mathcal{V}} \rho(\vec{r}\,')\,d^3r' \, \frac{\vec{r} - \vec{r}\,'}{|\vec{r} - \vec{r}\,'|^3} \, , \tag{3.4}$$

where d^3r' represents an infinitesimal volume element located at point $\vec{r}\,'$.

Exercise 3.2 Show that, for $\rho(\vec{r}\,') = q\,\delta(\vec{r}\,' - \vec{r}_o)$, Eq. (3.4) reduces to Eq. (3.3). Thus, the charge density of a point charge is accurately described by a three-dimensional Dirac-delta function.

3.2.1.3 Coulomb's law: differential form

Any well-behaved[1] vector field is uniquely specified by its divergence and curl[2]. We will soon show that the curl of $\vec{E}$ in electrostatics is zero. Hence, the $\vec{E}$-field in electrostatics is determined by its divergence alone. To find the divergence of sources of a given $\vec{E}$-field, we take divergence of $\vec{E}$, we take the divergence of both sides of Eq. (3.4). Moving the divergence operation inside the integral [3] and using the identity

$$\vec{\nabla} \cdot \frac{\vec{r} - \vec{r}\,'}{|\vec{r} - \vec{r}\,'|^3} = 4\pi\,\delta(\vec{r} - \vec{r}\,'),$$

[1]That is a continuous field with continuous first partial derivatives

[2]This is known as the Helmholtz theorem in vector analysis.

[3]This is justified because the divergence operator involves differentiation with respect to the components of $\vec{r}$, while the integration is over $\vec{r}\,'$

we get

$$\vec{\nabla} \cdot \vec{E} = \frac{\rho}{\epsilon_o} \ . \tag{3.5}$$

This is known as the Coulomb's law in differential form and is one of the four equations that comprise Maxwell's Equations of electrodynamics.

Exercise 3.3 Starting from Eq. (3.4), derive Eq. (3.5).

Example 3.1 $\vec{E}$-field due to a charged disk
Determine the electric field on the axis of a circular disk of radius R, carrying an azimuthally symmetric surface charge density $\eta = \eta_o r/R$, where r measures the distance from the axis of the circle, which we denote as the z-axis.

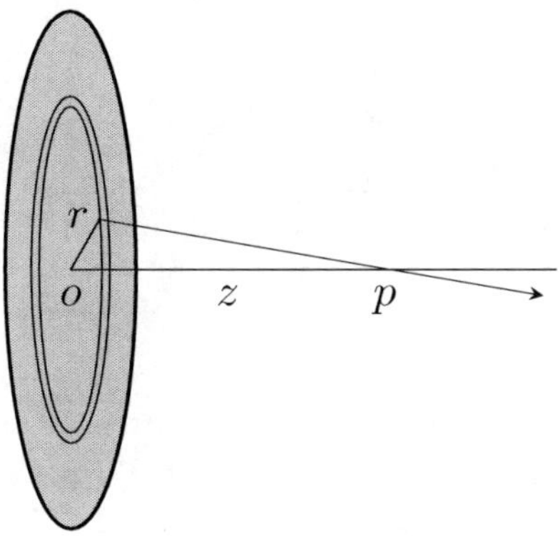

FIGURE 3.1
A uniformly charged circular disk.

Consider two circles of radii r and $r + dr$ on the disk, as shown in Fig. 3.1. The area between the two circles is a strip of width dr and length $2\pi r$. A small segment of the strip, that subtends an infinitesimal angle $d\varphi$, carries a charge $dq = \eta\, dA = \eta\, r\, d\varphi\, dr = (\eta_o r/R)\; r\, d\varphi\, dr$. The magnitude of the electric field dE, due to charge dq, at point p is given by

$$dE = \frac{1}{4\pi\epsilon_o}\, \frac{dq}{|r^2 + z^2|^2} = \frac{1}{4\pi\epsilon_o}\, \frac{\eta_o r^2\, d\varphi\, dr}{R\,|r^2 + z^2|^2}.$$

By symmetry, only the z-component of the electric field will be non-zero. Hence,

$$dE_z = dE\ \frac{z}{\sqrt{|r^2 + z^2|}} = \frac{1}{4\pi\epsilon_o}\, \frac{z\, r\, d\varphi\, dr\, \eta_o r/R}{(r^2 + z^2)^{3/2}} \ .$$

Hence,

$$E_z = \frac{z\,\eta_o}{4\pi\epsilon_o R}\int_0^{2\pi} d\varphi \int_0^R dr\,\frac{r^2}{(r^2+z^2)^{3/2}} \tag{3.6}$$

$$= \frac{2\pi z\,\eta_o}{4\pi\epsilon_o R}\left(-\frac{R}{\sqrt{R^2+z^2}} + \sinh^{-1}\left(\frac{R}{z}\right)\right) . \tag{3.7}$$

If we expand the expand the above expression for $z >> R$ and keeping only the first non-zero term, we get

$$E_z \cong \frac{2\pi z\,\eta_o}{4\pi\epsilon_o R}\frac{1}{3}\left(\frac{R^3}{z^3}\right) = \frac{1}{4\pi\epsilon_o}\frac{2\pi R^2\,\eta_o/3}{z^2} . \tag{3.8}$$

I.e., from very far away, the disk looks like a point charge with a total charge $Q = 2\pi R^2\,\eta_o/3$.

Exercise 3.4
(a) Starting from Eq. (3.7), derive Eq. (3.8).

(b) Integrating the surface charge density η given above, show that the total charge Q on the disk is indeed equal to $2\pi R^2\,\eta_o/3$.

3.2.2 Gauss' law

The divergence theorem connects the volume integration of the divergence of a vector field to the total flux of the field out of a closed surface $\mathcal{S}$ that is boundary to the volume; i.e.,

$$\int_{\mathcal{V}} \vec{\nabla}\cdot\vec{A}\;\mathrm{d}^3r = \int_{\mathcal{S}} \vec{A}\cdot\hat{n}\;\mathrm{dS} . \tag{3.9}$$

Using the above theorem we can write Eq. (3.5) as

$$\int_{\mathcal{S}} \vec{E}\cdot\hat{n}\;\mathrm{dS} = \frac{q_{\text{encl.}}}{\epsilon_o} , \tag{3.10}$$

where $\mathcal{S}$ is the surface that is boundary to the volume $\mathcal{V}$, and $q_{\text{encl.}}$ represents the net charge within the surface $\mathcal{S}$. As we will show though examples, Gauss' Law is specially effective in determining the electric field for systems with symmetric charge distribution and boundary conditions.

Example 3.2 $\vec{E}$-field: a spherically symmetric case
A charge Q is distributed uniformly within a sphere of radius R centered at the origin. Determine the electric field inside and outside of the sphere.

The charge density in this case can be written as

$$\rho(r,\theta,\varphi) = \begin{cases} \frac{3Q}{4\pi R^3} & r < R \\ 0 & r \geq R \end{cases} \tag{3.11}$$

Since the charge distribution is spherically symmetric, and because the divergence of the electric field is proportional to the charge density which is nonzero, the electric field must be radial; i.e., it cannot have θ or φ components.

To determine the electric field at any point, we divide the space into two parts: (1) inside of the sphere, $r < R$, and (2) outside of the sphere, $r > R$.

To determine the field inside the sphere, we imagine a spherical Gaussian surface of radius $r\,(r < R)$ that is concentric with the charge distribution. The total electric flux on this surface would be $4\pi r^2\, E_r$, where E_r is the radial component of the electrical field at the Gaussian surface. Then, from Eq (3.10), we get

$$4\pi r^2\, E_r = \frac{q_{\text{encl.}}}{\epsilon_o} = \left(4\pi r^3/3\right)\, \rho/\epsilon_o = \frac{Q\, r^3}{\epsilon_o\, R^3} \, ,$$

or

$$\vec{E}_{\text{in}} = \frac{\rho\, \vec{r}}{3\epsilon_o} = \frac{Q\, \vec{r}}{4\pi\epsilon_o\, R^3} \, . \tag{3.12}$$

Thus, inside the sphere, the field increases linearly with the radius.

For the region outside the sphere, we imagine a Gaussian surface of radius $r\,(r > R)$. Since this surface encloses the entire charge Q, from Eq (3.10), we get

$$4\pi r^2\, E_r = \frac{q_{\text{encl.}}}{\epsilon_o} = \frac{Q}{\epsilon_o} \, ,$$

and hence

$$\vec{E}_{\text{out}} = \frac{Q\, \vec{r}}{4\pi\epsilon_o\, r^3} \, . \tag{3.13}$$

If the sphere were to be shrunk to a point charge, as discussed before, the charge density would be given by the following delta-function distribution

$$\rho(r, \theta, \varphi) = Q\, \delta(\vec{r}) = \frac{Q}{4\pi r^2}\, \delta(r) \quad .$$

Exercise 3.5 Show that Eqs. (3.5), (3.12), and (3.13) correctly reproduce the charge density inside and outside of a uniformly charged sphere.

Example 3.3 Energy stored in a uniformly charged sphere
Using the expression $u_{\text{E}} = \frac{1}{2}\epsilon_o E^2$ for the energy density in electric field, determine the total energy stored in a uniform spherical charge distribution

of radius R. From Ex. (3.2), the energy stored in the field would be given by

$$
\begin{aligned}
U &= \int_0^R r^2 dr\, d\Omega \left(\frac{1}{2}\epsilon_o \vec{E}_{\text{in}}^2\right) + \int_R^\infty r^2 dr\, d\Omega \left(\frac{1}{2}\epsilon_o \vec{E}_{\text{out}}^2\right) \\
&= \frac{1}{2} 4\pi\epsilon_o \left(\frac{Q}{4\pi\epsilon_o}\right)^2 \left[\frac{1}{R^6}\int_0^R r^4 dr \;+\; \int_R^\infty \frac{1}{r^2} dr\right] \\
&= \frac{3}{5}\frac{Q^2}{4\pi\epsilon_o R} \, .
\end{aligned}
\tag{3.14}
$$

Clearly, if the charge distribution were to be compressed into a sphere of smaller radius, the total energy would increase. Note that, instead of computing the energy stored in the charged configuration, we have computed the energy stored in the resulting electric field.

3.2.3 Electrostatic potential

We now show that, for static distribution of charges, the electric field can be expressed as the gradient of a scalar function $V(\vec{r})$, known as the *electrostatic potential.* For many charge distributions, it is easier to first find the scalar potential and then derive the electric field. We begin with the vector identity

$$
\vec{\nabla} \frac{1}{|\vec{r}-\vec{r}\,'|} = -\frac{\vec{r}-\vec{r}\,'}{|\vec{r}-\vec{r}\,'|^3} \, , \tag{3.15}
$$

where the gradient operation $\vec{\nabla}$ involves differentiation with respect to the components of vector $\vec{r}$ and treats the components of vector $\vec{r}\,'$ as constants. The electric field can be written as

$$
\begin{aligned}
\vec{E}(\vec{r}) &= \frac{1}{4\pi\epsilon_o}\int_{\mathcal{V}} \rho(\vec{r}\,') \frac{\vec{r}-\vec{r}\,'}{|\vec{r}-\vec{r}\,'|^3}\, d^3r' \\
&= -\frac{1}{4\pi\epsilon_o}\int_{\mathcal{V}} \rho(\vec{r}\,')\, \vec{\nabla} \frac{1}{|\vec{r}-\vec{r}\,'|}\, d^3r' \\
&= -\frac{1}{4\pi\epsilon_o}\vec{\nabla}\int_{\mathcal{V}} \frac{\rho(\vec{r}\,')}{|\vec{r}-\vec{r}\,'|}\, d^3r' \, ,
\end{aligned}
\tag{3.16}
$$

where ρ is the charge density. We are allowed to pull the gradient operation out of the integral as it involves differentiation with respect to coordinates (x, y, z), while the integration is with respect to variable (x', y', z'). Thus, for a given charge distribution ρ, the electric field can be written as

$$
\vec{E}(\vec{r}) = -\vec{\nabla}\, V(\vec{r}) \, , \tag{3.17}
$$

where

$$
V(\vec{r}) = \frac{1}{4\pi\epsilon_o}\int_{\mathcal{V}} \frac{\rho(\vec{r}\,')}{|\vec{r}-\vec{r}\,'|}\, d^3r' \, , \tag{3.18}
$$

up to a constant term C which can be determined from the boundary conditions. Substituting the expression for $\vec{E}$ in Eq. (3.5), we find

$$\vec{\nabla} \cdot \vec{E} = -\vec{\nabla} \cdot \vec{\nabla} V(\vec{r}) = \frac{\rho}{\epsilon_o} . \tag{3.19}$$

I.e., electrostatic potential satisfies Poisson's equation

$$\vec{\nabla}^2 V(\vec{r}) = -\frac{\rho}{\epsilon_o} . \tag{3.20}$$

Exercise 3.6 Show that, for a set of point charges q_i, situated at points $\vec{r}_i{}'$, the Eq. (3.18) reduces to

$$V(\vec{r}) = \frac{1}{4\pi\epsilon_o} \sum_i \frac{q_i}{|\vec{r} - \vec{r}_i{}'|} .$$

Exercise 3.7 Using $\frac{dV(\vec{r})}{dr} = -E_r$ and $\lim_{r\to\infty} V(r) \to 0$, determine the potential function inside and outside of a uniformly charged sphere discussed in Sec. (3.3).

Exercise 3.8 Employing an alternate form of the energy density in an electrostatic field: $u_{\mathrm{E}} = \frac{1}{2}\rho V(\vec{r})$, compute the total electrostatic energy stored in the charged sphere discussed in Sec. (3.3). However, in this case, the integration only needs to be over the part of the space, where the charge density is nonzero and not the entire space.

3.2.3.1 Uniqueness of the solution of Poisson's equation

In this section, we will show that if boundary conditions on an enclosing surface are specified, then solution to Eq. (3.20) is unique.

Proof: Boundary conditions can be in specified in several ways:

1. Dirichlet boundary condition – potential is given at every point on the surface, or
2. Neumann boundary condition – normal component of electric field ($E_n = -\vec{\nabla} V \cdot \hat{n}$) is specified at every point of the surface, or
3. Cauchy boundary condition – both the potential and the field E_n are specified on the surface.

Assume that two distinct solutions Φ_1 and Φ_2 exist that satisfy Eq. (3.20) and one of the three boundary conditions given above. Define a function $\Phi_D = \Phi_1 - \Phi_2$. Our objective is to show that Φ_D is a constant. From Eq. (3.20), we find that Φ_D satisfies $\vec{\nabla}^2 \Phi_D(\vec{r}) = 0$. Additionally, since Φ_1 and

Φ_2 satisfy identical boundary conditions on the enclosing surface, we must have $\Phi_D = (\vec{\nabla}\Phi_D) \cdot \hat{n} = 0$ on the surface. Now, let us consider the vector identity $\vec{\nabla} \cdot (\Phi_D \vec{\nabla}\Phi_D) = \Phi_D \vec{\nabla}^2 \Phi_D + |\vec{\nabla}\Phi_D|^2$. Integrating $\vec{\nabla} \cdot (\Phi_D \vec{\nabla}\Phi_D)$ over the space occupied by the charge density and using the divergence theorem, we get

$$\begin{aligned} \int \vec{\nabla} \cdot (\Phi_D \vec{\nabla}\Phi_D)\, d^3r' &= \oint \Phi_D (\vec{\nabla}\Phi_D) \cdot \hat{n}\, dS \\ \int \Phi_D \vec{\nabla}^2 \Phi_D\, d^3r' + \int |\vec{\nabla}\Phi_D|^2\, d^3r' &= \oint \Phi_D (\vec{\nabla}\Phi_D) \cdot \hat{n}\, dS \\ \int |\vec{\nabla}\Phi_D|^2\, d^3r' &= 0\,. \end{aligned} \tag{3.21}$$

But $|\vec{\nabla}\Phi_D|^2 \geq 0$, hence, for its volume integral to be zero, we must have $\vec{\nabla}\Phi_D = 0$. I.e., $\Phi_D = \Phi_1 - \Phi_2 = C$, a constant. For Dirichlet and Cauchy boundary conditions, since Φ_1 and Φ_2 are equal at the surface, this constant must be zero.

Thus, if somehow we can find one solution for a problem with a given charge density and boundary conditions, the uniqueness theorem tells us that we have identified the one and the only solution that exists for that problem.

Example 3.4 Image method

Determine the potential at all points $z > 0$ due to a point charge q placed at $(0, 0, z_0)$, where $z_0 > 0$. The planar surface $z = 0$ is maintained at zero potential.

Fig. 3.2 shows a hemisphere of radius R (very large) that contains a charge q, situated at point $(0, 0, z_0)$. We want to determine the potential inside this sphere, with the radius of the sphere $R \to \infty$. The potential on all bounding surface to this region of interest (which we will denote as region $\mathcal{R}_1$) is zero.

To solve this problem, we consider an alternate problem that consists of two charges: q situated at point $(0, 0, z_0)$ and $-q$ situated at point $(0, 0, -z_0)$. This combination of charges would create a zero potential on the surface $z = 0$. On the spherical surface $(R, \theta < \pi/2, \varphi)$ the potential would be zero as well, provided $R \to \infty$. In other words, this alternate problem has exactly the same boundary conditions and charge density for region $\mathcal{R}_1$, as the original problem. Thus, from uniqueness, the potential in Region $\mathcal{R}_1$ due to the alternate problem would be exactly the same as the original problem, which is given by

$$V(x, y, z) = \frac{q}{4\pi\epsilon_0}\left(\frac{1}{\sqrt{x^2 + y^2 + (z - z_0)^2}} - \frac{1}{\sqrt{x^2 + y^2 + (z + z_0)^2}}\right). \tag{3.22}$$

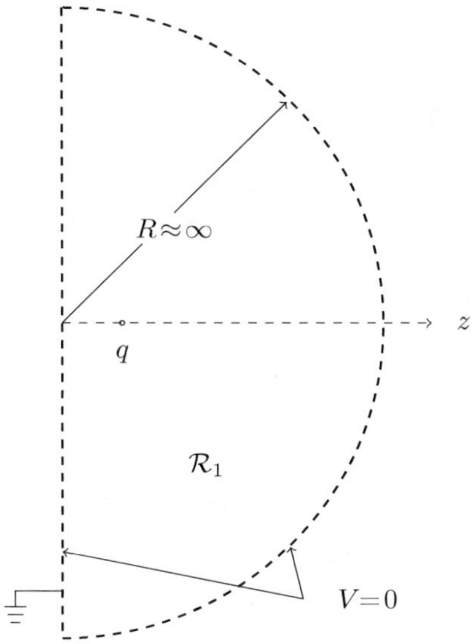

FIGURE 3.2
The region $\mathcal{R}_1$ includes a charge q situated at point $(0, 0, z_0)$, and the potential on all bounding surfaces are zero.

3.2.4 Charge conservation and continuity equation

The charge conservation is one of the sacrosanct principles of physics that states that the total electric charge of an isolated system is conserved. For any arbitrary volume, the change in total charge within any closed surface over a time interval Δt is always equal in magnitude, and opposite in sign, to the total charge that leaves that surface in the same time interval. This implies that the rate of change in total charge within a closed surface is equal in magnitude, and opposite in sign, to the total current out of the surface. For any closed surface $\mathcal{S}$, this principle can be expressed as

$$\frac{d}{dt} q_{\text{encl.}} = \frac{d}{dt} \int_{\mathcal{V}} \rho(\vec{r})\, d^3r = \int_{\mathcal{V}} \frac{\partial \rho(\vec{r})}{\partial t}\, d^3r = -\oint_{\mathcal{S}} \vec{J} \cdot \hat{n}\, \mathrm{dS} \; , \tag{3.23}$$

where $\vec{J} = \rho\vec{v}$ is the current density. Its surface integration over any area element gives the total current flowing across the surface. Since Eq. (3.23) is valid for any arbitrary volume $\mathcal{V}$, using the divergence theorem, we obtain

$$\frac{\partial \rho}{\partial t} + \vec{\nabla} \cdot \vec{J} = 0 \; . \tag{3.24}$$

Eq. (3.24) is the differential form of the charge conservation condition and is known as the *continuity equation.*

Exercise 3.9 Starting from Eq. (3.23), derive Eq. (3.24).

Example 3.5 Continuity equation for the steady-state case
For the steady state case; i.e., $\frac{\partial \rho}{\partial t} = 0$, show that current passing through any cross-section of a wire are the same.

Since for a steady-state $\frac{\partial \rho}{\partial t} = 0$, we have $\vec{\nabla} \cdot \vec{J} = 0$. For a segment of a wire, this implies that

$$\oint_S \vec{J} \cdot \hat{n}\, \mathrm{dS} = \int_{\text{sides}} \vec{J} \cdot \hat{n}\, \mathrm{dS} + \int_{\text{end 1}} \vec{J} \cdot \hat{n}\, \mathrm{dS} + \int_{\text{end 2}} \vec{J} \cdot \hat{n}\, \mathrm{dS} = 0\ .$$

Since charges cannot escape from the sides of the wire, we have

$$\int_{\text{sides}} \vec{J} \cdot \hat{n}\, \mathrm{dS} = 0\ .$$

This implies that the net current going out of the two sides equals zero, i.e., the current coming in from one end (a cross-section) must be equal to the current going out of the other end.

3.2.5 Biot-Savart law and Ampère's circuital law

The magnetic field $\vec{B}$ is created by moving charges. The $\vec{B}$-field lines due to a current element of length $d\vec{\ell}$ forms closed circle around the current segment, as shown in Fig. 3.3. A way to remember the direction of the $\vec{B}$-field is

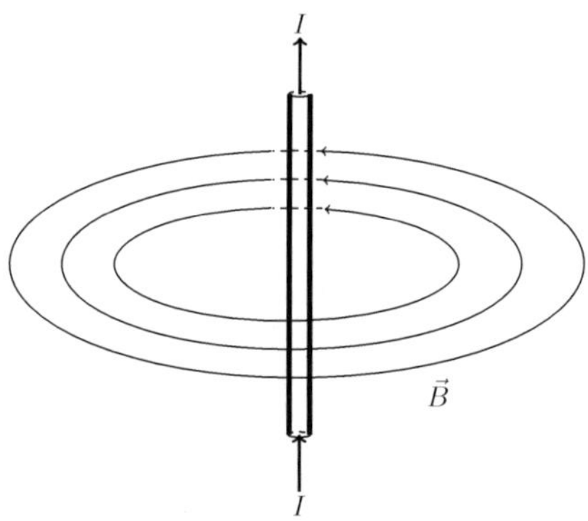

FIGURE 3.3
Contours of magnetic field around a linear current element.

to visualize grasping a linear current element with your right-hand thumb pointing along the direction of the current, then the $\vec{B}$ would curl around the current segment as your fingers around the wire – this is known as the *Ampère's circuital law.* A quantitative description of the field generated is given by *Biot-Savart law*

$$\mathrm{d}\vec{B}(r) = \frac{\mu_o}{4\pi}\,\frac{i\,\mathrm{d}\vec{\ell}\times\vec{r}}{r^3}\,, \tag{3.25}$$

where the line element $\mathrm{d}\vec{\ell}$ points along the current i, $\vec{r}$ points from the elemental current to the field point, and μ_o is a constant called the *magnetic permeability of vacuum*, whose value in SI units is $4\pi \times 10^{-7}\,TmA^{-1}$. It is important to note that $\mu_o\epsilon_o\,c^2 = 1$, where c is the speed of light in vacuum, which indicates a close relationship between electricity, magnetism, and light.

Exercise 3.10 By integrating the expression in Eq. (3.25), show that the magnetic field due a linear current element of length ℓ, carrying current i, at a point on its bisecting plane is given by

$$\vec{B} = \frac{\mu_o\,i}{2\pi\,r}\,\frac{\ell}{\sqrt{4r^2+\ell^2}}\,\hat{\varphi}\,, \tag{3.26}$$

where r is the distance from the mid-point of the current element to the field point, and $\hat{\varphi}$ is the unit vector along the azimuthal angle φ assuming z-axis to point along the direction of the current.

For an infinitely long linear current, the magnetic field is given by

$$\vec{B} = \frac{\mu_o\,i}{2\pi\,r}\,\hat{\varphi}\,. \tag{3.27}$$

Example 3.6 Magnetic field due to a closed loop
We will compute the $\vec{B}$-field at the center of a current carrying circular loop of radius R. Assume that the current appears counterclockwise, when seen from above, as shown in Fig. 3.4. The circle lies on the x-y plane with the origin at the center O, and the unit vector $\hat{k}$ points out of the plane of the paper. Let us now consider an infinitesimal section of the loop given by $\mathrm{d}\vec{\ell} = R\,\mathrm{d}\varphi\,\hat{e}_\varphi$. Per Eq. (3.25), the magnetic field $\mathrm{d}\vec{B}$ generated by this small segment of length $R\mathrm{d}\varphi$ at the center is

$$\begin{aligned}\vec{B} &= \frac{\mu_o\,i}{4\pi}\oint\frac{(R\,\mathrm{d}\varphi\,\hat{\varphi})\times(-R\,\hat{r})}{R^3}\\ &= \frac{\mu_o\,i}{4\pi}\oint\frac{(-\hat{\varphi}\times\hat{r})}{R}\,\mathrm{d}\varphi = \frac{\mu_o\,i}{4\pi}\,\frac{\hat{k}}{R}\oint\mathrm{d}\varphi = \frac{\mu_o\,i}{2R}\,\hat{k}\,,\end{aligned} \tag{3.28}$$

where unit vectors $\left(\hat{r},\hat{\varphi},\hat{k}\right)$ form a right-handed triad for the cylindrical coordinates. Remember that the symbol $\odot$ denotes a vector pointing out of the paper, as is shown at point O, and $\otimes$ denotes a vector going into the page.

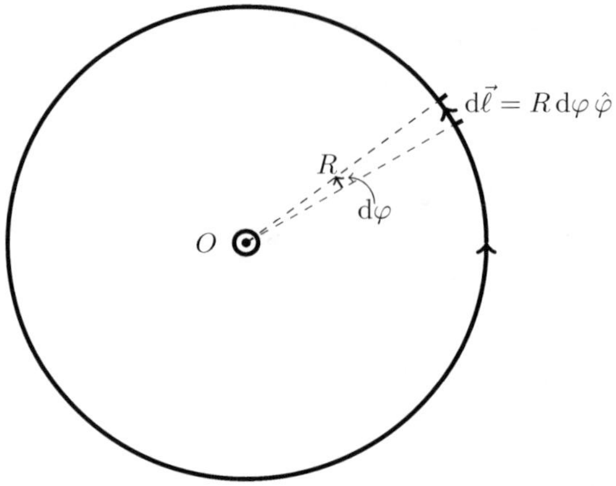

FIGURE 3.4
Magnetic field at the center of circular current.

It is worth noting that the direction of the magnetic field at any point inside the circle is given by $\hat{k}$; i.e., pointing out of the page, and at any point outside of the circle is along $-\hat{k}$.

3.2.6 Biot-Savart law in differential form

We will now generalize Biot-Savart law, Eq. (3.25), for an arbitrary current distribution. Integrating Eq. (3.25), the magnetic field at point $\vec{r}$ is given by

$$\vec{B}(\vec{r}) = \frac{\mu_o}{4\pi} \oint \frac{i\, d\vec{\ell}' \times (\vec{r} - \vec{r}\,')}{|\vec{r} - \vec{r}\,'|^3} , \tag{3.29}$$

where $d\vec{\ell}'$ denotes a current element situated at point $\vec{r}\,'$. For a wire with area of cross-section da, replacing i by $J(\vec{r}\,')\, da$, and then $J(\vec{r}\,')\, da\, d\vec{\ell}'$ by $\vec{J}(\vec{r}\,')\, d^3r'$, we get

$$\vec{B}(\vec{r}) = \frac{\mu_o}{4\pi} \int \frac{\vec{J}(\vec{r}\,') \times (\vec{r} - \vec{r}\,')}{|\vec{r} - \vec{r}\,'|^3}\, d^3r' , \tag{3.30}$$

where we defined $\vec{J}(\vec{r}\,')$ to have the same direction as $d\vec{\ell}'$. Taking curl of both sides of Eq. (3.30), we obtain

$$\vec{\nabla} \times \vec{B}(\vec{r}) = \mu_o\, \vec{J}(\vec{r}) . \tag{3.31}$$

Note that the curl operation is a differentiation with respect to the variable $\vec{r}$ and can be taken inside the integral on the right. Using the vector identity

$$\vec{\nabla} \times (\vec{A} \times \vec{B}) = \vec{A}(\vec{\nabla} \cdot \vec{B}) - \vec{B}(\vec{\nabla} \cdot \vec{A}) + (\vec{B} \cdot \vec{\nabla})\vec{A} - (\vec{A} \cdot \vec{\nabla})\vec{B} ,$$

we get

$$\vec{\nabla} \times \vec{B} = \frac{\mu_o}{4\pi} \int d^3r' \, \vec{J} \left[\vec{\nabla} \cdot \left(\frac{\vec{r} - \vec{r}\,'}{|\vec{r} - \vec{r}\,'|^3} \right) \right] - \frac{\mu_o}{4\pi} \int d^3r' \left[\vec{J} \cdot \vec{\nabla} \right] \frac{\vec{r} - \vec{r}\,'}{|\vec{r} - \vec{r}\,'|^3}.$$

Note that all terms with differential operators acting on $\vec{J}(\vec{r}\,')$ have disappeared. With the use of $\vec{\nabla} \cdot \frac{\vec{r}-\vec{r}\,'}{|\vec{r}-\vec{r}\,'|^3} = 4\pi\,\delta(\vec{r} - \vec{r}\,')$, the first integral on the right collapses to $\mu_o \vec{J}(\vec{r})$. The second integral on the right vanishes for a steady current distribution ($\vec{\nabla} \cdot \vec{J} = 0$) that is spatially localized, which means there exists a surface S that encloses all currents and hence $\oint_S \vec{J} \cdot \hat{n}\, dS = 0$. For more details, see [18]. Integrating Eq. (3.31) over an area, we get

$$\begin{aligned} \int \vec{\nabla} \times \vec{B}(\vec{r}) \cdot \hat{n}\, dS &= \mu_o \int \vec{J}(\vec{r}) \cdot \hat{n}\, dS \\ \oint \vec{B} \cdot d\vec{\ell} &= \mu_o I_{\text{enclosed}} \, . \end{aligned} \tag{3.32}$$

Eq. (3.32) is known as Ampere's Circuital Law. We will soon generalize Eq. (3.31) for time-varying currents and fields.

3.2.7 Divergence of magnetic field – monopole density

In Eq. (3.31) we determined the curl of the magnetic field and related it to current density. Since a vector field is completely determined by its curl and its divergence, we now compute $\vec{\nabla} \cdot \vec{B}$. We will now show that $\vec{B}$ is a solenoidal field; i.e., $\vec{\nabla} \cdot \vec{B} = 0$. Taking the divergence of both sides of Eq. (3.30) and using the vector identity $\vec{\nabla} \cdot \left(\vec{C} \times \vec{D}\right) = \vec{D} \cdot (\vec{\nabla} \times \vec{C}) - \vec{C} \cdot (\vec{\nabla} \times \vec{D})$, we get

$$\begin{aligned} \vec{\nabla} \cdot \vec{B}(\vec{r}) &= -\frac{\mu_o}{4\pi} \int \vec{J}(\vec{r}\,') \cdot \left(\vec{\nabla} \times \frac{(\vec{r} - \vec{r}\,')}{|\vec{r} - \vec{r}\,'|^3} \right) d^3r' \\ \vec{\nabla} \cdot \vec{B}(\vec{r}) &= 0 \, , \end{aligned} \tag{3.33}$$

where we have used $\vec{\nabla} \times \vec{J}(\vec{r}\,') = 0$, and $\frac{(\vec{r}-\vec{r}\,')}{|\vec{r}-\vec{r}\,'|^3} = -\vec{\nabla} \frac{1}{|\vec{r}-\vec{r}\,'|}$. The Eq. (3.33), one of the four Maxwell's equations, indicates that there are no monopoles; i.e., magnetic charges, and hence the surface integral $\oint \vec{B} \cdot d\vec{S}$ must vanish for all closed surfaces.

Example 3.7 Magnetic fields have neither beginnings nor ends
Show that, unlike electrical flux, two surfaces bounded by a common closed contour must have the same magnetic flux through them at any time.

Consider two surfaces, S_1 and S_2, that have as common boundary a closed curve $\mathcal{C}$. For simplicity, we assume that these surfaces do not intersect at any point other than the points on curve $\mathcal{C}$. We denote the volume between the two surfaces by $\mathcal{V}$. Now, using both the divergence theorem and $\vec{\nabla} \cdot \vec{B} = 0$, we obtain

$$\int_{\mathcal{V}} \vec{\nabla} \cdot \vec{B}\, d^3r = \oint \vec{B} \cdot \hat{n}\, \mathrm{dS} = \int_{\mathrm{S}_1} \vec{B} \cdot \hat{n}_1'\, \mathrm{dS} + \int_{\mathrm{S}_2} \vec{B} \cdot \hat{n}_2'\, \mathrm{dS} = 0 , \tag{3.34}$$

where $\hat{n}_1'$ and $\hat{n}_2'$ are unit vectors that are normal to surfaces S_1 and S_2 respectively, and point outward from $\mathcal{V}$. Defining vectors $\hat{n}_1 = -\hat{n}_1'$ and $\hat{n}_2 = \hat{n}_2'$, the Eq. (3.34) now reads

$$\int_{\mathrm{S}_1} \vec{B} \cdot \hat{n}_1\, \mathrm{dS} = \int_{\mathrm{S}_2} \vec{B} \cdot \hat{n}_2\, \mathrm{dS} .$$

3.2.8 Ohm's law

Ohm's law provides an approximate relationship between an applied electric field and the resulting current density in a medium. For a low electric field, most conducting materials exhibit a linear relationship between the current density and the applied field; i.e.,

$$\vec{J} = \sigma_c \vec{E} , \tag{3.35}$$

Here σ_c is the electric conductivity, which is constant for a homogeneous material. For a wire of uniform cross-section a, length, ℓ and potential difference ΔV across the wire, the above equation can be written in scalar form as

$$\begin{aligned} I/a &= \sigma_c\, \Delta V/\ell, \quad \text{or} \\ \Delta V &= I\, R \end{aligned} \tag{3.36}$$

where $R = \ell/\sigma_c a$ is known as the resistance of the wire. Both equations $\vec{J} = \sigma_c \vec{E}$ and $\Delta V = I\, R$ are known as Ohm's law.

4

Electrodynamics

4.1 Introduction

As we saw in the last chapter, charge and current distributions generate electrical and magnetic fields respectively. If these source distributions do not vary with time, the two fields remain distinct. However, when sources change with time, the $\vec{E}$ and $\vec{B}$ fields become intertwined.

In this chapter we will learn about time-varying electric and magnetic fields and the Maxwell's equations that relate them.

4.2 Electrodynamics

4.2.1 Faraday's law – electromagnetic induction

Extensive investigations by Faraday and Lenz, among others, revealed that that changing magnetic fields generates currents in a closed loop. These observations can be succinctly expressed by

$$\mathcal{E} = \oint_{C} \vec{E} \cdot \mathrm{d}\vec{\ell} = -\frac{d\Phi_m}{dt} , \tag{4.1}$$

where $\Phi_m = \int \vec{B} \cdot \hat{n} \, \mathrm{d}S$ denotes the magnetic flux passing through the closed loop $\mathcal{C}$, and $\mathcal{E}$ is the induced emf generated in the loop.

In this section, through examples and exercises, we will establish the relationship between the orientation of the emf and the direction of the normal unit vector $\hat{n}$ to a surface bounded by the contour $\mathcal{C}$. We start with a brief presentation of Biot-Savart law that describes the magnetic field produced by a current element, and compute the magnetic field for several current distributions. It is interesting to note that (see Eq. (4.1)), in the presence of a changing flux, $\oint_{\mathcal{C}} \vec{E} \cdot \mathrm{d}\vec{\ell} \neq 0$, and hence, in this case, the electric field is not a conservative vector field.

DOI: 10.1201/9781003174967-4

Example 4.1 Changing flux and induced emf
A circular loop of radius R is placed perpendicular to a uniform magnetic field that is varying with time. Its magnitude increases with time at a rate β. Compute the emf $\mathcal{E}$ generated in the loop.

Considering the loop to be in the x-y plane, the magnetic field is given by $\vec{B} = -B_o(t)\,\hat{k}$, and its rate of change, $\partial\vec{B}/\partial t$, is $-\beta\,\hat{k}$. Let us imagine an arbitrary surface that has the above circular loop as its boundary, and the normal vector at all points has a component that is in the $-\hat{k}$ direction. The flux through this surface would be exactly the same that would be through a planar surface containing the loop[1], which is given by

$$\Phi_m = \int \vec{B}\cdot\hat{n}\,\mathrm{d}S = \int \left(-B_o(t)\,\hat{k}\right)\cdot\left(-\hat{k}\,\mathrm{d}S\right) = \pi\,R^2\,B_o(t)\ .$$

From Faraday's law, the emf is then given by

$$\mathcal{E} = -\,\frac{d\Phi_m}{dt} \;=\; -\,\pi\,R^2\,\beta\ . \tag{4.2}$$

We now discuss the implication of the negative sign. Since Φ_m was computed with respect to a normal that was pointed in the $-\hat{k}$ direction, an induced emf that generates a current in a right-handed manner with respect to the normal would be considered positive. In this case, if we point our right-hand thumb along $-\hat{k}$, an induced current along the fingers would be considered positive. The sign for the emf generating this current is chosen using the same convention. Hence, the sign of emf $\mathcal{E} \;=\; -\,\pi\,R^2\,\beta$ indicates that the induced emf will be counterclockwise, as seen from above.

Lenz's law provides another path to arrive at the same result. As you may have seen before, Lenz's law states that if the flux through a loop is increasing, the induced emf will oppose it, and *vice versa.* For our example, the flux directed into the loop is increasing, with the magnetic field pointed towards $-\hat{k}$, and hence the induced emf will be such that it would produce magnetic field directed towards $\hat{k}$. Such a current must be counterclockwise, as seen from above.

Example 4.2 Electric field induced by changing magnetic field
Assume that we have an electromagnet that is in shape of a doughnut, with a thin slice taken out. The two circular poles that are close enough so that the magnetic field between the poles could be considered uniform. If the strength of the magnetic field changes with time, an electric field would be generated in the space between the poles. Compute this induced electric field.

[1] See Exercise (3.7).

Assume the z-axis to be along the centers of the two poles, and the magnetic field to be given by $B_o(t)\,\hat{k}$. From Faraday's law, we have

$$\begin{aligned}\oint_{\mathcal{C}} \vec{E}\cdot \mathrm{d}\vec{\ell} &= 2\pi r\, E_\varphi = -\frac{d\Phi_m}{dt} = -\dot{B}_\mathrm{o}\, \pi r^2 \\ E_\varphi &= -\frac{1}{2}\dot{B}_\mathrm{o}\, r\ ,\end{aligned} \tag{4.3}$$

where r measures distance from the z-axis, and φ is the azimuthal angle. Thus, Faraday's law not only gave us the magnitude of the induced electric field, it also gave us its direction.

4.2.2 Faraday's law in differential form

The Faraday's law

$$\mathcal{E} \equiv \oint_{\mathcal{C}} \vec{E}\cdot \mathrm{d}\vec{\ell} = -\frac{d\Phi_m}{dt}\ , \tag{4.4}$$

is valid for any closed loop. Using the identity $\oint_{\mathcal{C}} \vec{A}\cdot \mathrm{d}\vec{\ell} = \int_\mathrm{S} \left(\vec{\nabla}\times\vec{A}\right)\cdot \hat{n}\,\mathrm{dS}$, where S is a fixed arbitrary surface bounded by the curve $\mathcal{C}$, we can write Eq. (4.4) as

$$\begin{aligned}\int_\mathrm{S} \vec{\nabla}\times\vec{E}\cdot\hat{n}\,\mathrm{dS} &= -\frac{d}{dt}\int_\mathrm{S}\vec{B}\cdot\hat{n}\,\mathrm{dS}\ , \\ &= -\int_\mathrm{S}\frac{\partial\vec{B}}{\partial t}\cdot\hat{n}\,\mathrm{dS}\ .\end{aligned}$$

Since this is true for an arbitrary surface, we must have

$$\vec{\nabla}\times\vec{E} = -\frac{\partial\vec{B}}{\partial t}\ . \tag{4.5}$$

This is one of the four Maxwell's equation of electrodynamics.

4.2.3 Ampere-Maxwell equation

So far, we have seen that a changing magnetic field generates an electric field. Using a generalization of the Ampere's law, we will now show that the reverse is true as well; i.e., a changing electric field is also a source of magnetic field.

We start with the simple electrical circuit shown in Fig. 4.1. When the switch $\mathcal{S}$ is closed, a current starts flowing in counterclockwise direction.

If we imagine a circular Amperian loop enclosing the wire as illustrated in Fig. 4.2, the net current piercing through the planar surface of the wire would then be related to the line integral of the magnetic field around the

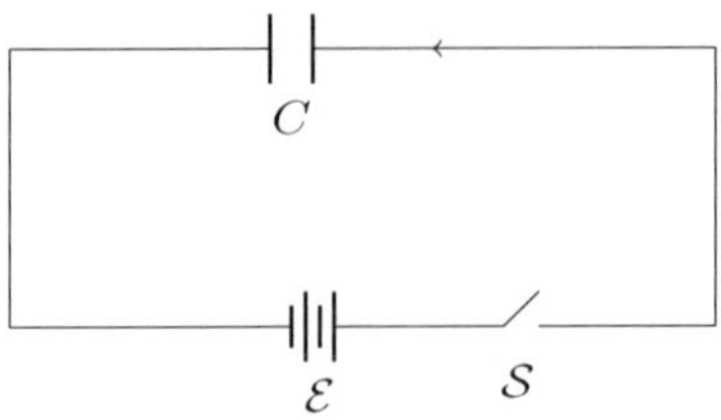

FIGURE 4.1
A simple circuit involving a capacitor C, a battery $\mathcal{E}$, and a switch $\mathcal{S}$.

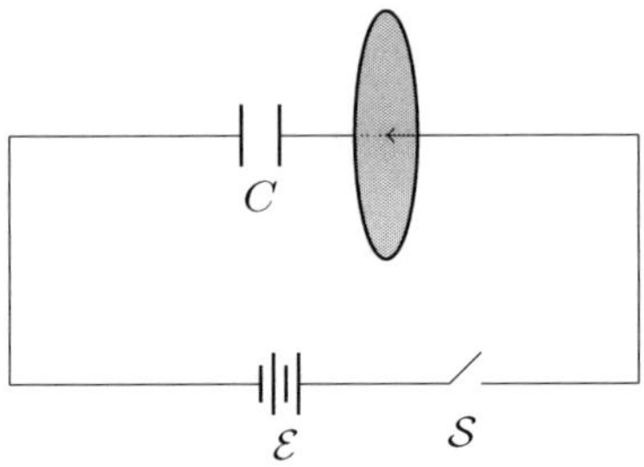

FIGURE 4.2
A circular planar surface, perpendicular to the wire, bounded by a circular Amperian loop.

loop computed in a right-handed manner. We thus have

$$\oint \vec{B} \cdot d\vec{\ell} = \mu_o \, I_{\text{enclosed}} \ .$$

As stated in Eq. (3.32), I_{enclosed} is the surface integral of the current density over the surface bounded by the loop. However, if we stretch the surface enclosed by the loop to include one of the plates of the capacitor, as shown in Fig. 4.3, we face an apparent inconsistency. Since no current seems to pierce through the surface in Fig. 4.3, we now have $I_{\text{enclosed}} = 0$. This paradox can also be seen at the level of the differential form of the Ampere's law, Eq. (3.31). If we take the divergence of both sides of Eq. (3.31), we get $\vec{\nabla} \cdot \vec{J} = 0$, because the divergence of the curl vanishes. From the equation of continuity,

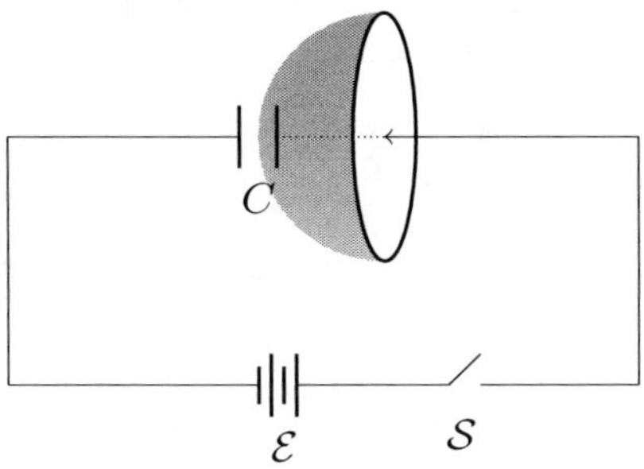

FIGURE 4.3
The planar surface of Fig. 4.2 deformed into a hemisphere to enclose one of the plates of the capacitor C. The deformed surface is still bounded by the circular loop of Fig. 4.2.

Eq. (3.24), $\vec{\nabla} \cdot \vec{J}$ is not zero in general if the charge density changes with time. Using Coulomb's law, we can write Eq. (3.24) as

$$\vec{\nabla} \cdot \vec{J} + \frac{\partial \rho}{\partial t} = \vec{\nabla} \cdot \vec{J} + \epsilon_o \frac{\partial}{\partial t} \vec{\nabla} \cdot \vec{E} = \vec{\nabla} \cdot \left(\vec{J} + \epsilon_o \frac{\partial \vec{E}}{\partial t} \right) = 0 \,. \tag{4.6}$$

If we replace $\vec{J}$ by $\vec{J} + \epsilon_o \frac{\partial \vec{E}}{\partial t}$ in Eq. (3.31), the above inconsistency disappears. The term $\epsilon_o \frac{\partial \vec{E}}{\partial t}$ was added by James C. Maxwell and is known as the *displacement current density*. Thus, Maxwell used the continuity equation to provide a pathway to generalize Biot-Savart's and Ampere's laws to address the more general cases of electrodynamics. The generalized Ampere-Maxwell equation is then given by

$$\vec{\nabla} \times \vec{B}(\vec{r}) = \mu_o \vec{J}(\vec{r}) + \mu_o \epsilon_o \frac{\partial \vec{E}}{\partial t} \,. \tag{4.7}$$

4.3 Maxwell's equations

In Chapter 3 we presented four equations that govern electricity and magnetism when the sources and boundary conditions were time independent. In this chapter, we saw their generalizations for the time varying situations.

These four equations, written below, are Maxwell's equations:

$$\vec{\nabla} \cdot \vec{E}(\vec{r}) = \frac{\rho}{\epsilon_o} \,, \qquad \text{Colomb's Law} \tag{4.8}$$

$$\vec{\nabla} \cdot \vec{B}(\vec{r}) = 0 \,, \qquad \text{No magnetic monopole} \tag{4.9}$$

$$\vec{\nabla} \times \vec{E}(\vec{r}) = -\frac{\partial \vec{B}(\vec{r})}{\partial t} \,, \qquad \text{Faraday's Law} \tag{4.10}$$

$$\vec{\nabla} \times \vec{B}(\vec{r}) = \mu_o \left(\vec{J}(\vec{r}) + \epsilon_o \frac{\partial \vec{E}(\vec{r})}{\partial t} \right) \,, \qquad \text{Ampere-Maxwell's Law} \tag{4.11}$$

Together with the continuity equation, which is an expression of the charge conservation, Maxwell's equations form the pillars of electrodynamics.

4.4 Maxwell's equations in vacuum

In vacuum; i.e., in the absence of charge and current densities, Maxwell's equations reduce to

$$\vec{\nabla} \cdot \vec{E} = 0, \qquad \vec{\nabla} \times \vec{E} = -\frac{\partial \vec{B}}{\partial t} \,,$$

$$\vec{\nabla} \cdot \vec{B} = 0 \,, \qquad \vec{\nabla} \times \vec{B} = \mu_o \epsilon_o \frac{\partial \vec{E}}{\partial t} \,. \tag{4.12}$$

Since, for any general vector field $\vec{D}$, $\vec{\nabla} \times \vec{\nabla} \times \vec{D} = \vec{\nabla}(\vec{\nabla} \cdot \vec{D}) - \vec{\nabla}^2 \vec{D}$, from Eqs. (4.12), we get the wave equations

$$\vec{\nabla}^2 \vec{E} - \mu_o \epsilon_o \frac{\partial^2 \vec{E}}{\partial t^2} = 0, \qquad \vec{\nabla}^2 \vec{B} - \mu_o \epsilon_o \frac{\partial^2 \vec{B}}{\partial t^2} = 0 \,. \tag{4.13}$$

These equations represent a wave moving with a speed $c = \frac{1}{\sqrt{\mu_o \epsilon_o}} = 3 \times 10^8 \, \mathrm{m\,s^{-1}}$, which shows a deep connection between light and electromagnetism.

4.5 Problems

4.1 A circular ring of radius R caries a uniform linear charge density $\lambda = Q/2\pi R$.

(a) Determine the electric field on the axis of the ring and show that this expression reduces to that of a point charge for $z >> R$, where z-axis is assumed to be along the axis of symmetry with the origin at the center of the disk.

(b) Assume that a charge $-Q$ of mass m is placed on the axis, at a distance z from the center of the ring. Considering the ring to be fixed in space, show that for *small* values of z, the charge will undergo simple harmonic motion. Find the frequency of the oscillatory motion.

4.2 Following example 3.1 from Chapter 3,

(a) determine the electric field on the axis of a uniformly charged disk of radius R and surface charge density σ and show that this expression reduces to that of a point charge for $z >> R$;

(b) determine the electric field due to an infinite sheet of uniform charge density σ.

4.3 From Problem 4.2 above, using $\vec{E} = -\vec{\nabla}V$,

(a) determine the potential on the axis of the disk;

(b) using the result from part (a) find the potential off the axis.

4.4 Consider a spherical shell of radius R with a uniform surface charge distribution $\sigma = Q/4\pi R^2$.

(a) Using Gauss' law, determine the electric field inside and outside of the sphere.

(b) Is the normal component of the electrical field continuous across the surface of the sphere?

4.5 Consider a cylindrically symmetric current distribution given by $\vec{J}(r, \varphi, z) = rJ_0/R\ \Theta(1 - r/R)\,\hat{z}$, where Θ is the Heaviside step function.

(a) Determine the total current carried by this current distribution.

(b) Using Ampere's law, find the magnetic field inside and outside of the cylinder.

(c) Determine the difference in the normal and tangential components of the magnetic field across a cylindrical surface or radius R.

4.6 Consider a cylindrical conducting shell of radius R with a uniform surface current density $\vec{K} = i/2\pi R\,\hat{z}$, where i represents the total current flowing along the wire.

(a) Using Ampere's law, determine the magnetic field inside and outside of the cylinder.

(b) Are the normal and tangential components of the magnetic field continuous across the surface of the cylinder? Justify your answer.

4.7 Assume that the electrical field in empty space (i.e., no free charge or current density $\vec{J}$) is given by $\vec{E}(x, y, z, t) = E_0\,\hat{\imath}\,\sin(kz - \omega t)$. From $\vec{\nabla} \times \vec{E} + \frac{\partial \vec{B}}{\partial t} = 0$, determine the magnetic field $\vec{B}(x, y, z, t)$.

4.8 Show that a free charge ρ_o placed on a material of conductivity σ_c will decay exponentially with time.

5

Conservation Principles

5.1 Introduction

Conservation principles play a very important role in mechanics. This is also true in electromagnetism, and in physics in general. In this chapter we will study the energy and momentum carried by electric and magnetic fields and their conservation laws.

5.2 Energy in electromagnetic field

In this section we start with the work done by the electromagnetic field on the moving charges and then use conservation principles to derive the flow of electromagnetic energy across surfaces. We first compute work done by electromagnetic field on moving charges.

5.2.1 Lorentz force and work done on moving charges

The force experienced by a charged particle in electromagnetic field is given by the Lorentz force formula:

$$\vec{F} = q\left(\vec{E} + \vec{v} \times \vec{B}\right) , \tag{5.1}$$

where q is the charge, and $\vec{v}$ is the velocity of the particle. The work done on a particle undergoing a displacement $\delta\vec{r} = \vec{v}\,\delta t$ is

$$\delta W = \vec{F} \cdot \vec{v}\,\delta t = q\left(\vec{E} + \vec{v} \times \vec{B}\right) \cdot \vec{v}\,\delta t = q\vec{E} \cdot \vec{v}\,\delta t . \tag{5.2}$$

Hence, the work done on a set of charges q_i during an infinitesimal time interval δt can be written as

$$\delta W = \sum_i q_i \vec{v}_i \cdot \vec{E}_i\,\delta t , \tag{5.3}$$

where $\vec{E}_i$ is the electric field experienced by the charge q_i.

DOI: 10.1201/9781003174967-5

For a continuous distribution of charge, the work done on the charge enclosed within a small volume ΔV, small enough so that the charge density ρ can be assumed constant, becomes

$$\delta W = \rho\,\Delta V\,\vec{v}\cdot\vec{E}\,\delta t\ . \tag{5.4}$$

Then, the power expended per unit volume by the field is

$$\mathcal{P} = \rho\vec{v}\cdot\vec{E} = \vec{J}\cdot\vec{E}\ . \tag{5.5}$$

If these charges moved through a wire, Eq. (5.5) would give rise to Joule heating.

Exercise 5.1 Use Eq. (5.5), and Ohm's law, to show that the power expended in a wire of resistance $\mathcal{R}$, carrying a current i, is $i^2\mathcal{R}$.

5.2.2 Poynting vector and the energy flow

The work done by the electromagnetic field on charges, Eq. (5.5), results in the reduction in energy of the field. To see this, we note that from Maxwell's equations (4.10) and (4.11), we have

$$\vec{\nabla}\times\vec{E}(\vec{r}) = -\frac{\partial\vec{B}(\vec{r})}{\partial t}\ , \quad \text{and} \quad \vec{\nabla}\times\vec{B}(\vec{r}) = \mu_o\left(\vec{J}(\vec{r}) + \epsilon_o\,\frac{\partial\,\vec{E}(\vec{r})}{\partial t}\right)\ .$$

Taking the scalar product of the first equation with $\vec{B}$ and of the second equation with $\vec{E}$, and then subtracting one from another, we get

$$\vec{E}\cdot\left(\vec{\nabla}\times\vec{B}\right) - \vec{B}\cdot\left(\vec{\nabla}\times\vec{E}\right) = \mu_o\left(\vec{J}\cdot\vec{E} + \epsilon_o\,\vec{E}\cdot\frac{\partial\,\vec{E}}{\partial t}\right) + \vec{B}\cdot\frac{\partial\vec{B}}{\partial t}\ .$$

Now we use the identity

$$\vec{\nabla}\cdot\left(\vec{E}\times\vec{B}\right) = \vec{B}\cdot\left(\vec{\nabla}\times\vec{E}\right) - \vec{E}\cdot\left(\vec{\nabla}\times\vec{B}\right)\ ,$$

to obtain

$$\vec{J}\cdot\vec{E} + \frac{1}{\mu_o}\vec{\nabla}\cdot\left(\vec{E}\times\vec{B}\right) = -\frac{\partial}{\partial t}\left(\frac{\epsilon_o}{2}\,\vec{E}^2 + \frac{1}{2\,\mu_o}\vec{B}^2\right)\ . \tag{5.6}$$

The right-hand side of Eq. (5.6) is the rate at which energy is lost by the electromagnetic field. The term $\vec{J}\cdot\vec{E}$ on the left-hand side shows the rate at which the energy is expended within a unit volume due to the work done on the charges, see Eq. (5.5). From energy conservation, the second term on the left must be interpreted as the energy leaving the unit volume per unit time. The vector $\vec{S} = \frac{1}{\mu_o}\left(\vec{E}\times\vec{B}\right)$ that describes this power flow is known as *Poynting vector*.

Exercise 5.2 Show that in *steady state*, Eq. (5.6) reduces to

$$\int \vec{J} \cdot \vec{E}\, d\tau = -\oint \frac{1}{\mu_o}\left(\vec{E} \times \vec{B}\right) \cdot \hat{n}\, dS = 0 \,, \tag{5.7}$$

where $d\tau$ and $\hat{n}\, dS$ represent infinitesimal volume and surface elements respectively. Thus, in *steady state*, the energy for work done on charges within any volume must flow in from outside of its surface.

Exercise 5.3 For a wire of radius r, length ℓ and a potential difference of $\Delta\Phi$ across it, show that the power flow from its surface into the wire must be equal to the product of the current i and the potential difference $\Delta\Phi$.

Example 5.1 Charging of a capacitor
For the circuit shown below, Fig. 5.1, determine power as a function of time, and the total energy, that is fed to the capacitor as the charge on the capacitor reaches its steady-state value.

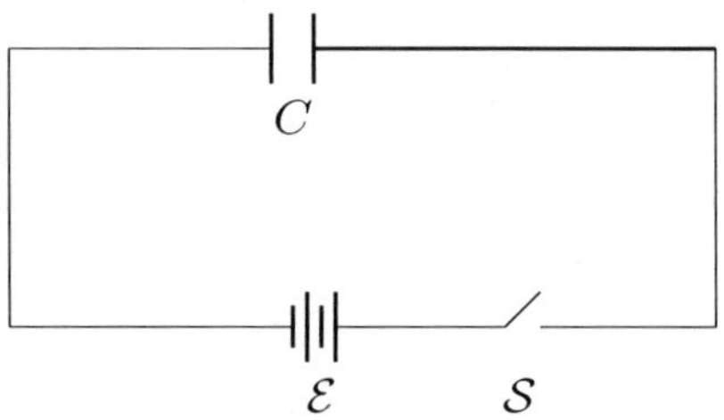

FIGURE 5.1
As switch $\mathcal{S}$ is pressed, a current grows in the wire and electric field builds up in the capacitor. The battery has an internal resistance $\mathcal{R}$.

As switch $\mathcal{S}$ is closed, the charge q on the capacitor increases as

$$q = C\mathcal{E}\left(1 - e^{-\frac{t}{\tau}}\right) \,,$$

where τ is the *time constant* of the circuit, equal to $\mathcal{R}C$. The current dq/dt in the circuit is

$$i = \frac{\mathcal{E}}{\mathcal{R}} e^{-\frac{t}{\tau}} \,.$$

Consequently, the electric and magnetic fields between the plates of the capacitor are given by

$$\vec{E} = \frac{\mathcal{E}}{d}\left(1 - e^{-\frac{t}{\tau}}\right) \quad \text{and} \quad \vec{B} = \frac{\mu_o\, \mathcal{E}}{2\pi r \mathcal{R}} e^{-\frac{t}{\tau}} \,,$$

where r measures the distance from the axis of the capacitor plates and d is the distance between the plates. Assuming the plates of the capacitor to be circular and of radius R, we calculate the rate of energy flow into the space between the capacitor plates by integrating the Poynting vector $\vec{S} = \frac{1}{\mu_o}\left(\vec{E}\times\vec{B}\right)$ over the cylindrical surface of area $2\pi Rd$, enclosing the capacitor. This yields:

$$\oint \vec{S}\cdot d\vec{a} = \frac{1}{\mu_o}\left[\frac{\mathcal{E}}{d}\left(1-e^{-\frac{t}{\tau}}\right)\cdot\frac{\mu_o\,\mathcal{E}}{(2\pi R)\mathcal{R}}e^{-\frac{t}{\tau}}\right](2\pi Rd) = \frac{\mathcal{E}^2}{\mathcal{R}}\left(1-e^{-\frac{t}{\tau}}\right)e^{-\frac{t}{\tau}} .$$

The total energy stored in the capacitor is then given by

$$U = \int_0^\infty \frac{\mathcal{E}^2}{\mathcal{R}}\left(1-e^{-\frac{t}{\tau}}\right)e^{-\frac{t}{\tau}}\,dt = \frac{1}{2}\mathcal{C}\mathcal{E}^2 .$$

5.3 Momentum density in electromagnetic field

In the previous section, we saw that the electromagnetic field carries energy. In particular, we saw that the work done by the electromagnetic field on charges and on current densities, together with the flow of energy out of the enclosing surface results in a decrease of the energy stored within that volume. We will now learn that a similar conservation principle also governs the momentum of the electromagnetic field.[1]

We begin with the Lorentz force formula for a point charge in electromagnetic field, Eq. (5.1):

$$\vec{F} = q\left(\vec{E}+\vec{v}\times\vec{B}\right) ,$$

which for a continuous charge distribution implies

$$\Delta\vec{F} = \left(\rho\vec{E}+\vec{J}\times\vec{B}\right)\Delta V ,$$

where, ΔV is a small volume with a charge density ρ. For a unit volume reduces to the force density

$$\vec{\mathcal{F}} = \rho\vec{E}+\vec{J}\times\vec{B} . \tag{5.8}$$

[1] In 1902, Ernest Fox Nichols and Gordon Ferrie Hull showed that electromagnetic waves carry momentum [35].

As we had done in the previous section, we express this density $\vec{\mathcal{F}}$ solely in terms of the fields by replacing ρ and $\vec{J}$ with $\epsilon_o \vec{\nabla} \cdot \vec{E}$ and $\left(\frac{1}{\mu_o} \vec{\nabla} \times \vec{B} - \epsilon_o \frac{\partial \vec{E}}{\partial t} \right)$, respectively. We write Eq. (5.8) as

$$\vec{\mathcal{F}} = \left(\epsilon_o \vec{\nabla} \cdot \vec{E} \right) \vec{E} + \left(\frac{1}{\mu_o} \vec{\nabla} \times \vec{B} - \epsilon_o \frac{\partial \vec{E}}{\partial t} \right) \times \vec{B} \,. \tag{5.9}$$

Substituting $\frac{\partial \vec{E}}{\partial t} \times \vec{B}$ for $\frac{\partial}{\partial t} \left(\vec{E} \times \vec{B} \right) - \vec{E} \times \frac{\partial \vec{B}}{\partial t}$, and $\frac{\partial \vec{B}}{\partial t}$ with $-\vec{\nabla} \times \vec{E}$, we get

$$\vec{\mathcal{F}} = \epsilon_o \left(\vec{\nabla} \cdot \vec{E} \right) \vec{E} + \epsilon_o \left(\vec{\nabla} \times \vec{E} \right) \times \vec{E} + \frac{1}{\mu_o} \left(\vec{\nabla} \times \vec{B} \right) \times \vec{B} - \epsilon_o \frac{\partial}{\partial t} \left(\vec{E} \times \vec{B} \right) \,.$$

Since $\vec{\nabla} \cdot \vec{B}$ is identically zero, we can bring a total symmetry between $\vec{E}$ and $\vec{B}$ by adding a term $\frac{1}{\mu_o} \left(\vec{\nabla} \cdot \vec{B} \right) \vec{B}$ to arrive at

$$\begin{aligned} \vec{\mathcal{F}} &= \epsilon_o \left(\vec{\nabla} \cdot \vec{E} \right) \vec{E} + \epsilon_o \left(\vec{\nabla} \times \vec{E} \right) \times \vec{E} \\ &+ \frac{1}{\mu_o} \left(\vec{\nabla} \cdot \vec{B} \right) \vec{B} + \frac{1}{\mu_o} \left(\vec{\nabla} \times \vec{B} \right) \times \vec{B} \\ &- \frac{1}{c^2} \frac{\partial \vec{S}}{\partial t} \,, \end{aligned} \tag{5.10}$$

where we have used $\epsilon_o \mu_o = \frac{1}{c^2}$ and $\vec{S} = \frac{1}{\mu_o} \left(\vec{E} \times \vec{B} \right)$. Employing the identity

$$\left(\vec{\nabla} \times \vec{E} \right) \vec{E} = \left(\vec{E} \cdot \vec{\nabla} \right) \vec{E} - \frac{1}{2} \vec{\nabla}(E^2) \,.$$

we replace $\left(\vec{\nabla} \times \vec{E} \right) \vec{E}$ and $\left(\vec{\nabla} \times \vec{B} \right) \vec{B}$ in Eq. (5.10), which yields

$$\begin{aligned} \vec{\mathcal{F}} &= \epsilon_o \left[\left(\vec{\nabla} \cdot \vec{E} \right) \vec{E} + \left(\vec{E} \cdot \vec{\nabla} \right) \vec{E} - \frac{1}{2} \vec{\nabla}(E^2) \right] \\ &+ \frac{1}{\mu_o} \left[\left(\vec{\nabla} \cdot \vec{B} \right) \vec{B} + \left(\vec{B} \cdot \vec{\nabla} \right) \vec{B} - \frac{1}{2} \vec{\nabla}(B^2) \right] \\ &- \frac{1}{c^2} \frac{\partial \vec{S}}{\partial t} \,, \end{aligned} \tag{5.11}$$

For the i-th component of $\vec{\mathcal{F}}$, we get

$$\begin{aligned} \mathcal{F}_i &= \epsilon_o \left[(\partial_j E_j)\, E_i + E_j\, (\partial_j E_i) - \frac{1}{2} \partial_i (E_j E_j) \right] \\ &+ \frac{1}{\mu_o} \left[(\partial_j B_j)\, B_i + B_j\, (\partial_j B_i) - \frac{1}{2} \partial_i (B_j B_j) \right] \\ &- \frac{1}{c^2} \frac{\partial S_i}{\partial t} \,. \end{aligned} \tag{5.12}$$

Exercise 5.4 Prove Eq. (5.12).

5.3.1 Electromagnetic energy-momentum tensor

The above equation for the force density can be written in a compact form as

$$\mathcal{F}_i \;=\; \sum_{j=1}^{3} \partial_j T_{ji} - \frac{1}{c^2}\frac{\partial S_i}{\partial t} \,, \tag{5.13}$$

where, T_{ij} is the electromagnetic energy-momentum tensor T_{ij}:

$$T_{ij} = \left[\epsilon_o\, E_i E_j + \frac{1}{\mu_o} B_i B_j - \frac{1}{2}\left(\epsilon_o\, E^2 + \frac{1}{\mu_o} B^2\right)\delta_{ij}\right] \,. \tag{5.14}$$

Due to our familiarity with vectors, it is instructive to express the energy-momentum tensor as a dyad, which behaves like a *product* of two vectors. We first summarize some of the properties of dyads, and will then express $\vec{\mathcal{F}}$ in terms of a dyad.

Exercise 5.5 In terms of field components E_x, E_y, E_z and B_x, B_y, B_z, write the components of T_{ij} in a matrix form.

5.3.1.1 Basics of a dyad

A dyad behaves like a juxtaposition of two vectors with very specific multiplicative properties. As an example, let us consider a dyad $\overleftrightarrow{bd}$ created by the juxtaposition of two vectors $\vec{b}$ and $\vec{d}$. When we take a scalar product of this dyad with a vector $\vec{c}$, the end result is a vector:

$$\overleftrightarrow{bd} \cdot \vec{c} = \vec{b}\left(\vec{d}\cdot\vec{c}\right) \quad \text{and} \quad \vec{c}\cdot\overleftrightarrow{bd} = \left(\vec{c}\cdot\vec{b}\right)\vec{d}\,.$$

Similarly, we can define vector products of a dyad with a vector that results in another dyad:

$$\overleftrightarrow{bd} \times \vec{c} = \overleftrightarrow{be}\,, \quad \text{where } \vec{e} = \vec{d}\times\vec{c}\,;$$

and

$$\vec{c}\times\overleftrightarrow{bd} = \overleftrightarrow{fd}, \quad \text{where } \vec{f} = \vec{c}\times\vec{b}\,.$$

A dyad in three-dimensional space has nine components. The ij-components of the dyad $\overleftrightarrow{bd}$ are given by $b_i d_j$.

We define the energy-momentum dyad $\overleftrightarrow{T}$ built from the components of the corresponding to the energy-momentum tensor:

$$\overleftrightarrow{T} = \sum_{i,j=1}^{3} T_{ij}\overleftrightarrow{e_i e_j}.$$

where $\hat{e}_i,\ \hat{e}_j$ are unit vectors.

Exercise 5.6 Show that $\hat{e}_i \cdot \overleftrightarrow{T} \cdot \hat{e}_j = T_{ij}$.

In terms of $\overleftrightarrow{T}$, Eq. (5.13) can be written as

$$\vec{\mathcal{F}} = \vec{\nabla} \cdot \overleftrightarrow{T} - \frac{1}{c^2}\frac{\partial \vec{S}}{\partial t} . \tag{5.15}$$

The force on a finite volume ΔV then becomes

$$\begin{aligned} \vec{F} &= \int_{\Delta V} \vec{\nabla} \cdot \overleftrightarrow{T}\, d\tau - \frac{1}{c^2}\int_{\Delta V} \frac{\partial \vec{S}}{\partial t}\, d\tau \\ &= \oint \overleftrightarrow{T} \cdot \hat{n}\, da - \frac{\partial}{\partial t}\int_{\Delta V} \frac{\vec{S}}{c^2}\, d\tau , \end{aligned} \tag{5.16}$$

where we assumed that the volume element ΔV does not change with time.

Eq. (5.16) is the statement of momentum conservation. Net force on the charges and current densities in volume ΔV must equal the inward flow of momentum per unit time through the enclosing surface, and the rate at which the net momentum decreases within the volume. This then identifies $\vec{S}/c^2$ as the momentum density of the electromagnetic field, and $\overleftrightarrow{T} \cdot \hat{n}$ as the inward momentum flux.

Example 5.2 Force due to the image charge
For the setup of example (3.4), show that the force on the infinite grounded conducting plate due to the point charge q is the same as the force acting on its image charge.

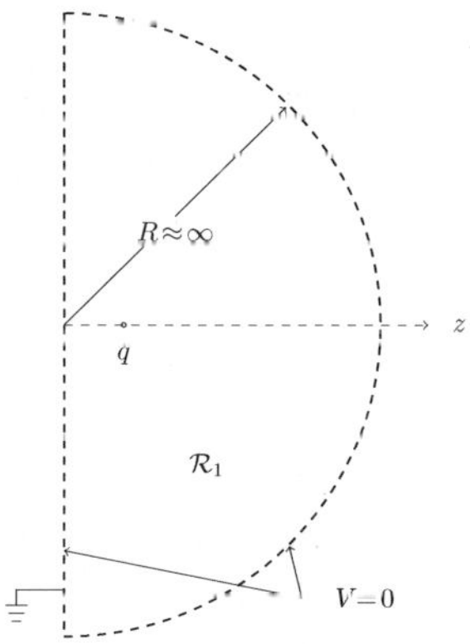

FIGURE 5.2
The region $\mathcal{R}_1$ includes a charge q situated at point $(0, 0, z_0)$. The conducting plane at $z = 0$ is grounded.

In this problem, we have a charge q situated on the z-axis, at a distance z_0 from a grounded conducting plane at $z = 0$. Per Eq. (3.22), the potential in cylindrical coordinates (r, ϕ, z), at any point $z > 0$ is given by

$$V(r, \phi, z) = \frac{q}{4\pi\epsilon_o}\left(\frac{1}{\sqrt{r^2 + (z - z_o)^2}} - \frac{1}{\sqrt{r^2 + (z + z_o)^2}}\right) . \tag{5.17}$$

From symmetry considerations, the force on the plane must be along the z-direction. Since the fields are not changing in time, from Eq. (5.16) we have

$$F_z = \oint \hat{z} \cdot \overleftrightarrow{T} \cdot \hat{n}\, da .$$

The term $\oint \hat{z} \cdot \overleftrightarrow{T} \cdot \hat{n}\, da$ describes the momentum being transferred into the volume per unit time. We will carry out this integral on the infinite plane $z = 0$, so that the integral captures the rate of momentum transfer from the positive towards the negative z. The area element on the plane in cylindrical coordinate is then given by $\hat{n}\, da = 2\pi\, r\, dr\, \hat{z}$. Thus, we have

$$F_z = \int_0^\infty T_{zz}\ (2\pi\, r\, dr) . \tag{5.18}$$

From Eq. (5.14), the component T_{zz} is given by

$$T_{zz} = \left[\epsilon_o\, E_z E_z - \frac{1}{2}\epsilon_o\, E^2\right] = \frac{1}{2}\epsilon_o\, E_z^2 . \tag{5.19}$$

From Eq. (5.17), the electric field $E_z = -\frac{\partial V}{\partial z}$ at z_{0+} is given by

$$E_z = -\frac{q z_o}{2\pi\epsilon_o\, (\rho^2 + {z_o}^2)^{3/2}} .$$

Hence, the energy-momentum tensor at the grounded plane, $z \to 0+$, is then given by

$$T_{zz} = \frac{q^2 z_o^2}{8\pi^2\epsilon_o\, (\rho^2 + z_o^2)^3} .$$

Substituting T_{zz} into Eq. (5.18), we get

$$F_z = \frac{q^2}{16\pi\epsilon_o z_o^2} = \frac{q^2}{4\pi\epsilon_o\, (2z_o)^2} ,$$

which is exactly what we expect the force to be on the image charge $-q$ situated at a distance $(2z_o)$.

5.4 Problems

5.1 Assume that the electrical and magnetic field in empty space (i.e., no free charge or current density $\vec{J}$) are given by $\vec{E}(x,y,z,t) = E_0\,\hat{\imath}\,\sin(kz-\omega t)$ and $\vec{B}(x,y,z,t) = B_0\,\hat{\jmath}\,\sin(kz-\omega t)$.

- Show that these fields are related by $\vec{\nabla} \times \vec{E} + \frac{\partial \vec{B}}{\partial t} = 0$
- Determine the time-varying Poynting vector $\vec{S}$.
- Determine time-averaged $\vec{S}$-field over one period.

5.2 Determine the force between two equal charges q_1 and q_2, that are a distance d apart. Hint: Integrate the energy-momentum tensor on a plane bisecting the line joining the two charges.

5.3 Compute the force per unit length between two infinitely long parallel wires, carrying equal and opposite currents I_1 and I_2, that are a distance d apart. Hint: Integrate the energy-momentum tensor on a plane symmetrically placed between the two wires.

5.4 Compute the force on the upper half of uniformly charged spherical shell of total charge Q, due to the lower half

(a) by integrating the energy-momentum tensor on an infinite horizontal plane bisecting the sphere, and

(b) by integrating the energy-momentum tensor on a hemispherical closed surface enclosing the upper half of the shell that uses the bisecting plane for closure.

5.5 Compute the force on the upper half of uniformly charged spherical ball of total charge Q, due to the lower half

(a) by integrating the energy-momentum tensor on an infinite horizontal plane bisecting the sphere, and

(b) by integrating the energy-momentum tensor on a hemispherical closed surface enclosing the upper half of the ball that uses the bisecting plane for closure.

6

Scalar and Vector Potentials

6.1 Introduction

In mechanics, we have seen that while forces are measurable quantities as they have direct effect on the dynamics of a system, potential energies are not. We can always add an arbitrary constant to the potential energy, without any measurable consequences. Since only the difference in potential energy between two configurations is meaningful, we have the freedom to assign a zero of the potential energy to a convenient configuration of the system, which often simplifies the analysis.

For conservative systems we can express the force as the gradient of the potential energy. Many times, it is easier to work with one scalar term, the potential energy, than with the three components of the force.

Similarly, the dynamics of electromagnetic field involves six fields, the components of $\vec{E}$ and $\vec{B}$-fields, which are constrained by two homogeneous Maxwell's equations: $\vec{\nabla} \cdot \vec{B} = 0$ and $\vec{\nabla} \times \vec{E} + \frac{\partial \vec{B}}{\partial t} = 0$. In this chapter we will express the field components by a scalar field and a vector field, that are unconstrained, and ensure that all constraints on $\vec{E}$ and $\vec{B}$-fields are automatically met.

6.2 Electrostatics

The electrostatic field is known to be conservative, i.e., the work done by the field on a charge is independent of the path, and only depends on the initial and final positions $\vec{r}_1$ and $\vec{r}_2$. Thus, we have

$$q \int_{\vec{r}_1 \ (\text{path}\, 1)}^{\vec{r}_2} \vec{E} \cdot d\vec{\ell} = q \int_{\vec{r}_1 \ (\text{path}\, 2)}^{\vec{r}_2} \vec{E} \cdot d\vec{\ell} \, .$$

The above statement can be written as

$$\int_{\vec{r}_1 \ (\text{path}\, 1)}^{\vec{r}_2} \vec{E} \cdot d\vec{\ell} + \int_{\vec{r}_2 \ (\text{path}\, 2)}^{\vec{r}_1} \vec{E} \cdot d\vec{\ell} \equiv \oint \vec{E} \cdot d\vec{\ell} = 0 \, .$$

DOI: 10.1201/9781003174967-6

Thus, to any point $\vec{r}$, we can define a potential function $V(\vec{r})$ given by

$$V(\vec{r}) = -\int_{\vec{r}_1}^{\vec{r}} \vec{E} \cdot d\vec{\ell}\,, \tag{6.1}$$

where we have arbitrarily chosen $V(\vec{r}_1) = 0$. In turn, the electric field can be written as

$$\vec{E}(\vec{r}) = -\vec{\nabla}\, V(\vec{r})\,. \tag{6.2}$$

Exercise 6.1 Using Eq. (6.1) derive Eq. (6.2).

Here we note a few things about the scalar potential.

1. The negative sign indicates that the potential $V(\vec{r})$ decreases along the direction of $\vec{E}$.
2. An electrostatic force depends on the electric field $\vec{E}$, and hence we can add a constant to the potential without having any measurable effect.
3. Eq. (6.2) shows that the three components of an electrostatic field are constrained by the condition $\vec{\nabla} \times \vec{E} = 0$, which is a consequence of the electric field being expressed as the gradient of the scalar potential.

Exercise 6.2 Determine the potential difference between points $\vec{r}_1$ and $\vec{r}_2$, due to a charge q situated at $\vec{r}_o$. Hint: *Since the integral $\int_{\vec{r}_1}^{\vec{r}_2} \vec{E} \cdot d\vec{\ell}$ is path independent, you may choose paths that are part of circles about $\vec{r}_o$, or are along radial lines originating from $\vec{r}_o$.*

Exercise 6.3 Because of the constraints, not any vector field can represent an electrostatic field $\vec{E}$; however you certainly can imagine an arbitrary electrostatic potential. Determine the electric field and charge density of the electrostatic potential $V = V_o\, e^{-\left(\frac{r}{r_o}\right)^2}$.

6.3 Magnetostatics

While for an electrostatic field we can write $\vec{E} = -\vec{\nabla}\, V$, this is not true for time varying fields. In general, the curl of electric fields is not zero. Therefore, they cannot always be written as the gradient of a scalar potential. As we know from the Faraday's law, in presence of a time-dependent magnetic field

$\vec{\nabla} \times \vec{E} = -\frac{\partial \vec{B}}{\partial t} \neq 0$. However, the $\vec{\nabla} \cdot \vec{B}$ is always zero, and hence we can express

$$\vec{B} = \vec{\nabla} \times \vec{A} \,, \tag{6.3}$$

where $\vec{A}$ is called the *vector potential* and plays a role similar to the scalar potential V. Just as we could add a constant to the scalar potential V, we can add the gradient of an arbitrary scalar field χ to $\vec{A}$ and not change the measurable field $\vec{B}$. This freedom of varying potentials V and $\vec{A}$ without changing the fields $\vec{E}$ and $\vec{B}$ is known as the *gauge transformation* .

In case of magnetostatics, neither $\vec{E}$ or $\vec{B}$ are time dependent, and hence we have

$$\vec{\nabla} \times \vec{B} = \mu_o \vec{J} \,. \tag{6.4}$$

Substituting $\vec{B} = \vec{\nabla} \times \vec{A}$ in the above equation, we get

$$\vec{\nabla} \left(\vec{\nabla} \cdot \vec{A} \right) - \vec{\nabla}^2 \vec{A} = \mu_o \vec{J} \,. \tag{6.5}$$

Given that we have the freedom of adding to $\vec{A}$ the gradient of an arbitrary function χ, we can choose a new vector

$$\vec{A}' = \vec{A} + \vec{\nabla} \chi \,, \tag{6.6}$$

such that $\vec{\nabla} \cdot \vec{A}' = 0$. Choosing an appropriate χ to make $\vec{\nabla} \cdot \vec{A}'$ identically zero is known as the *Coulomb gauge* . .

In Coulomb gauge, using Eq. (6.5), the vector potential is given by a familiar equation

$$\vec{\nabla}^2 \vec{A} = -\mu_o \vec{J} \,, \tag{6.7}$$

which for each component is a Poisson equation. For spatially bounded current distributions $\vec{J}$, the solution is

$$\vec{A}(\vec{r}) = \frac{\mu_o}{4\pi} \int \frac{\vec{J}(\vec{r}\,')}{|\vec{r} - \vec{r}\,'|} \, d^3\vec{r}\,' \,. \tag{6.8}$$

This process of choosing a particular function χ is called *gauge fixing* .

It is important to note that we can always find such a χ. For this we assume that the field $\vec{A}$ is bounded in space and it goes to zero at ∞, and its divergence is not zero. Let us set

$$\vec{\nabla} \cdot \vec{A} = \lambda \,,$$

where λ is a scalar function of the position vector $\vec{r}$. Taking the divergence of the both sides of $\vec{A}' = \vec{A} + \vec{\nabla} \chi$, we get

$$\vec{\nabla} \cdot \vec{A}' = \vec{\nabla} \cdot \vec{A} + \vec{\nabla}^2 \chi \;=\; \lambda + \vec{\nabla}^2 \chi \,.$$

By setting $\vec{\nabla} \cdot \vec{A}' = 0$, we obtain

$$\vec{\nabla}^2 \chi = -\lambda \,;$$

i.e., χ must obey the Poisson's equation, and hence we have

$$\chi(\vec{r}) = \frac{1}{4\pi} \int_{\mathcal{V}} \frac{\lambda(\vec{r}')}{|\vec{r} - \vec{r}'|} \, d^3r' \, . \tag{6.9}$$

Furthermore, as long as we know $\vec{\nabla} \cdot \vec{A}$ over a closed volume $\mathcal{V}$ including its enclosing surface, the solution of Eq. (6.9) is unique.

Example 6.1 Magnetic field due to a linear current
As an example, we will now consider the magnetostatic case of an infinitely long wire carrying a steady current I along the z-axis. The current density $\vec{J}$ and the magnetic field $\vec{B}$ are not bounded in space. However, due to the

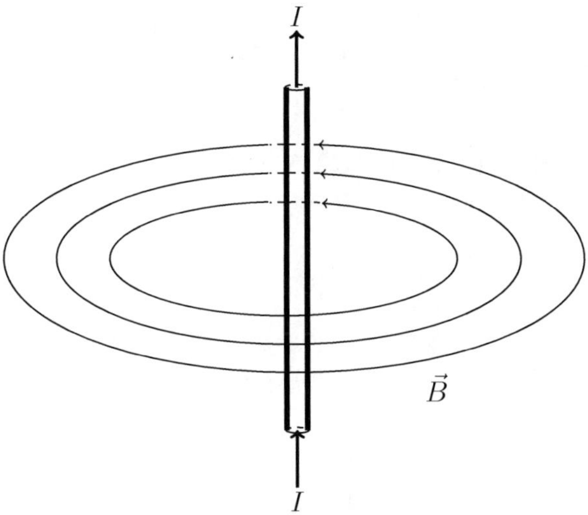

FIGURE 6.1
A linear current I flows along the z-axis, and the $\vec{B}$ is directed along $\hat{\varphi}$.

symmetry of the system, we can find the vector potential $\vec{A}$. Given that $\vec{B}$-field is circular, the vector potential is found to be

$$\vec{A} = -\left(\frac{\mu_0 I}{2\pi} \ln \frac{r}{r_0}\right) \hat{z} \; ; \quad r > r_0 . \tag{6.10}$$

Note that the direction of $\vec{A}$ is parallel to the current density $\vec{J}$, which should not be surprising due to Eq. (6.8).

Exercise 6.4 Derive Eq. (6.10).

Hint: From Example (3.10), we know that the field for this linear current is

$$\vec{B} = \frac{\mu_o I}{2\pi r}\hat{\varphi} \,. \tag{6.11}$$

Hence, find a vector field $A_r\hat{r} + A_\varphi\hat{\varphi} + A_z\hat{z}$ in cylindrical coordinate such that $\vec{\nabla}\times\vec{A} = \frac{\mu_o I}{2\pi r}\hat{\varphi}$.

Exercise 6.5 Show that the vector potential $\vec{A} = -\frac{\mu_o I}{2\pi}\ln\frac{r}{r_0}\hat{z}$ is in the Coulomb gauge; i.e., its divergence is zero.

6.4 Time-dependent fields

We now consider the case where both $\vec{E}$ and $\vec{B}$ are time dependent and obey the following homogeneous equations

$$\vec{\nabla}\cdot\vec{B} = 0 \quad ; \quad \vec{\nabla}\times\vec{E} + \frac{\partial\vec{B}}{\partial t} = 0 \,.$$

The first equation implies that we can write the magnetic field as

$$\vec{B} = \vec{\nabla}\times\vec{A} \,.$$

Then, from the second equation we have

$$\vec{\nabla}\times\vec{E} + \frac{\partial}{\partial t}\vec{\nabla}\times\vec{A} = \vec{\nabla}\times\left(\vec{E} + \frac{\partial\vec{A}}{\partial t}\right) = 0 \,.$$

Therefore, for a time-varying situation we can write the combination $\vec{E} + \frac{\partial\vec{A}}{\partial t}$ as the gradient of a scalar field. Hence, we set

$$\vec{E} + \frac{\partial\vec{A}}{\partial t} = -\vec{\nabla}V \,, \tag{6.12}$$

where V is a scalar potential. Thus, the electromagnetic fields $\vec{E}$ and $\vec{B}$ can be written in terms of the potentials V and $\vec{A}$ as

$$\vec{E} = -\frac{\partial\vec{A}}{\partial t} - \vec{\nabla}V \quad ; \quad \vec{B} = \vec{\nabla}\times\vec{A} \,. \tag{6.13}$$

The remaining two Maxwell's equations are:

$$\vec{\nabla}\cdot\vec{E}(\vec{r},t) = \frac{\rho(\vec{r},t)}{\epsilon_o} \quad ; \quad \vec{\nabla}\times\vec{B}(\vec{r},t) = \mu_o\left(\vec{J}(\vec{r},t) + \epsilon_o\frac{\partial\vec{E}(\vec{r},t)}{\partial t}\right) \,.$$

These equations are inhomogeneous, as they include source densities $\rho(\vec{r},t)$ and $\vec{J}(\vec{r},t)$.

Expressing the fields $\vec{E}$ and $\vec{B}$ in terms of potentials V and $\vec{A}$, we get the following equations:

$$\frac{\partial}{\partial t}\left(\vec{\nabla}\cdot\vec{A}\right)+\vec{\nabla}^2 V = -\frac{\rho}{\epsilon_o}\,, \tag{6.14}$$

$$-\vec{\nabla}\left(\vec{\nabla}\cdot\vec{A}+\mu_o\epsilon_o\frac{\partial V}{\partial t}\right)+\left(\vec{\nabla}^2\vec{A}-\mu_o\epsilon_o\frac{\partial^2\vec{A}}{\partial t^2}\right) = -\mu_o\vec{J}\,. \tag{6.15}$$

Thus, the problem of solving four Maxwell's equations to determine the six components of the fields $\vec{E}$ and $\vec{B}$ has been reduced to solving two equations, (6.14) and (6.15), and thus to finding four components for the potentials; one component for V and three for $\vec{A}$.

In order to investigate Eqs. (6.14) and (6.15), we first note that in classical electrodynamics only the fields $\vec{E}$ and $\vec{B}$ are measurable. The potentials V and $\vec{A}$ are mathematical artifacts which were created to make our analysis of $\vec{E}$ and $\vec{B}$ easier. They are not unique, and we have quite a bit of freedom in choosing them. In particular, since the magnetic field $\vec{B}$ is given by $\vec{B}=\vec{\nabla}\times\vec{A}$, we know that we can always define an equivalent vector potential $\vec{A}'$ by

$$\vec{A}'=\vec{A}+\nabla\chi\,,$$

which would produce the same magnetic field as the one produced by $\vec{A}$. However, unlike the time-independent case where the addition of χ would not affect $\vec{E}$, in this case we must modify the scalar potential V, in order to ensure that the electric field remains unchanged. This requires that

$$V\to V'=V-\frac{\partial\chi}{\partial t}\,.$$

Every non trivial choice of the function χ generates potentials V and $\vec{A}$ which could be specifically chosen to simplify our problem. This act of choosing a specific χ is called the *gauge fixing* and the fact that measurable fields $\vec{E}$ and $\vec{B}$ do not depend on a specific gauge fixing is known as *gauge invariance*. We will now see how we can choose an appropriate fixing of gauge to simplify the equations for potentials V and $\vec{A}$.

6.4.1 Coulomb gauge

In Coulomb gauge, we choose χ such that $\vec{\nabla}\cdot\vec{A}=0$. Then Eqs. (6.14) and (6.15) lead to

$$\vec{\nabla}^2 V(\vec{r},t) = -\frac{\rho(\vec{r},t)}{\epsilon_o}\,,$$

$$\mu_o\epsilon_o\vec{\nabla}\left(\frac{\partial V(\vec{r},t)}{\partial t}\right)+\left(\vec{\nabla}^2\vec{A}(\vec{r},t)-\mu_o\epsilon_o\frac{\partial^2\vec{A}(\vec{r},t)}{\partial t^2}\right)=\mu_o\vec{J}(\vec{r},t)\,.$$

The potential $V(\vec{r}, t)$ satisfies the Poisson equation and its solution at any time t is

$$V(\vec{r}, t) = \frac{1}{4\pi\epsilon_o} \int_{\mathcal{V}} \frac{\rho(\vec{r}\,', t)}{|\vec{r} - \vec{r}\,'|}\, d^3r' \,. \tag{6.16}$$

The fact that the same time t appears in both sides of the Eq. (6.16), can be disconcerting, because the equation doesn't seem to take into account the time required for a change in ρ to propagate in space. It shows that the potential $V(\vec{r}, t)$ responds instantaneously to a remote change in ρ. This would lead to an unphysical result if V alone would determine the electric field $\vec{E}$. But, since $\vec{E} = -\vec{\nabla} V - \frac{\partial \vec{A}}{\partial t}$, once the contribution of the $\vec{A}$-field is included, both $\vec{E}$ and $\vec{B}$ will show the appropriate time dependence.

6.4.2 Lorentz gauge

In Lorenz gauge, we set $\vec{\nabla} \cdot \vec{A} + \mu_o\epsilon_o \frac{\partial V}{\partial t} = 0$. From Eqs. (6.14) and (6.15), we then get

$$\vec{\nabla}^2 V - \mu_o\epsilon_o \frac{\partial^2 V}{\partial t^2} = -\frac{\rho}{\epsilon_o} \,, \tag{6.17}$$

$$\vec{\nabla}^2 \vec{A} - \mu_o\epsilon_o \frac{\partial^2 \vec{A}}{\partial t^2} = -\mu_o \vec{J} \,. \tag{6.18}$$

Thus, we see that both V and $\vec{A}$-fields obey inhomogeneous wave equations. There is great deal of symmetry between the two potentials in this gauge. We will see that this symmetry between equations for $V(\vec{r}, t)$ and $\vec{A}(\vec{r}, t)$ plays very important roles in special relativity.

Exercise 6.6 Given $V(\vec{r}, t)$ and $\vec{A}(\vec{r}, t)$, find a function χ such that the new potentials $V'(\vec{r}, t)$ and $\vec{A}'(\vec{r}, t)$ satisfy the Lorentz Gauge condition $\vec{\nabla} \cdot \vec{A}\,' + \mu_o\epsilon_o \frac{\partial V'}{\partial t} = 0$.

6.5 Problems

6.1 Prove that the vector field $\vec{A}$ given by Eq. (6.8) solves Eq. (6.7).

6.2 A conducting spherical shell of inner radius a_1 and outer radius a_2 has a charge Q at the center. Assuming the potential to be zero at the infinity, determine the potential at all points inside and outside of the shell.

6.3 An infinitely long conducting cylindrical shell of inner radius a_1 and outer radius a_2 has a line charge density λ_0 along its axis. Determine the potential at all points inside and outside of the shell. Note that the potential does not go to zero at infinity due the extended nature of the charge distribution.

6.4 An infinitely long cylindrical coaxial wire consists of an inner shell of radius a and outer shell of radius b. Determine the vector potential at all points inside and outside of the wire.

6.5 An infinite sheet of current, confined to the $x = 0$ plane is described by the current density $\vec{J} = K_0\,\hat{z}\,\delta(x)$. Determine $\vec{B}-$ and $\vec{A}-$fields in regions $x < 0$ and $x > 0$ for this current distribution.

6.6 Starting from Eq. (6.8),

(a) determine the vector potential at point $\vec{r}$ due to an elemental current $i\,d\vec{\ell}$ situated at the origin.

(b) Derive Eq. (6.10).

6.7 Explore the possibility that we could work with just only one potential.

(a) If you set the scalar potential to zero, what equation would the vector potential need to satisfy?

(b) If you set the vector potential to zero, what equation would scalar potential satisfy? Does the ability to set the vector potential to zero depend on the nature of the source densities?

6.8 Determine the electric and magnetic fields due to the following set of potentials and interpret corresponding results:

A. $$V(\vec{r},t) = 0 \text{ and } \vec{A}(\vec{r},t) = \frac{-Q_o}{4\pi\epsilon_o\,r^2}\,t\,\hat{r}\ ;$$

B. $$V(\vec{r},t) = V_0\,e^{-r/r_o} \text{ and } \vec{A}(\vec{r},t) = -\frac{\mu_o\,i}{2\pi}\ln\left(\frac{r}{r_0}\right)\hat{z} - \frac{V_0\,e^{-r/r_o}}{r_o}\,t\,\vec{r}\,,$$

where $\vec{r}$ points radially outward from the z-axis in the cylindrical coordinate system.

7

Radiation

7.1 Introduction

Stationary charges do not radiate. Hence, a charge moving with a constant velocity does not radiate either, as it could be viewed as a stationary charge in an inertial frame of reference tied to the particle. In this chapter we will learn about radiation from accelerating charges. In particular, we will study how time dependent charge and current densities generate electromagnetic fields and how they contribute to radiation.

7.2 Fields due to time varying charge and current distributions

In Chapter 5 we learned that the electromagnetic fields $\vec{E}$ and $\vec{B}$ could be generated from a scalar potential V and a vector potential $\vec{A}$:

$$\vec{E} = -\frac{\partial \vec{A}}{\partial t} - \vec{\nabla} V \quad \text{and} \quad \vec{B} = \vec{\nabla} \times \vec{A} \, .$$

Then, exploiting the gauge degree of freedom and choosing Lorentz gauge we found that the potentials obey the following two inhomogeneous wave equations:

$$\vec{\nabla}^2 V - \mu_o \epsilon_o \frac{\partial^2 V}{\partial t^2} = -\frac{\rho}{\epsilon_o} \, , \tag{7.1}$$

$$\vec{\nabla}^2 \vec{A} - \mu_o \epsilon_o \frac{\partial^2 \vec{A}}{\partial t^2} = -\mu_o \vec{J} \, . \tag{7.2}$$

For the time independent case, they reduce to

$$\vec{\nabla}^2 V = -\frac{\rho}{\epsilon_o} \quad , \quad \vec{\nabla}^2 \vec{A} - \mu_o \vec{J} \, . \tag{7.3}$$

As we have seen before, the solutions for Eq. (7.3) are

$$V(\vec{r}) = \frac{1}{4\pi\epsilon_o} \int_{\mathcal{V}} \frac{\rho(\vec{r}\,')}{|\vec{r} - \vec{r}\,'|} \, d^3 r' \quad , \quad \vec{A}(\vec{r}) = \frac{\mu_o}{4\pi} \int_{\mathcal{V}} \frac{\vec{J}(\vec{r}\,')}{|\vec{r} - \vec{r}\,'|} \, d^3 r' \, .$$

DOI: 10.1201/9781003174967-7

For the time dependent case, we need to take into account that any changes to charge and current density at point $\vec{r}'$ will induce a change in the potentials at point $\vec{r}$ with a time delay of $|\vec{r}-\vec{r}'|/c$. I.e., $V(\vec{r},t)$ and $\vec{A}(\vec{r},t)$ at time t should be related to the charge and current densities located at $\vec{r}'$ at an earlier time

$$t_r \equiv t - \frac{|\vec{r}-\vec{r}'|}{c} \ .$$

The time t_r is known as the . *retarded time* Then, the solutions to Eqs. (7.1-7.2) ought to be given by

$$V(\vec{r},t) = \frac{1}{4\pi\epsilon_o}\int_{\mathcal{V}} \frac{\rho(\vec{r}',t_r)}{|\vec{r}-\vec{r}'|}\, d^3r' \ , \quad \vec{A}(\vec{r},t) = \frac{\mu_o}{4\pi}\int_{\mathcal{V}} \frac{\vec{J}(\vec{r}',t_r)}{|\vec{r}-\vec{r}'|}\, d^3r' \ . \quad (7.4)$$

These potentials are called *retarded potentials.*

Even though we presented these solutions without proof, they are correct. They incorporate causality; i.e., the time for the interaction to travel from the source point $\vec{r}'$ to the field point $\vec{r}$ at the speed of light is explicitly built into these solutions. Using these expressions for the time dependent potentials, we will first investigate the radiation from sources that are localized in one place. Then, we will investigate the radiation from sources that are moving relative to the point of observation.

7.3 Radiation from localized sources

In this section we will study radiation far away from charge and current densities that are localized in space; i.e., they are contained within a finite volume of order d^3, where d represents the typical dimension of the distribution. The distance d is expected to be very small as compared to the distance r of the field point where the radiation is being observed. Since radiation is generated by accelerated charges, the charges must be in motion. For an oscillatory charge distribution with time period τ, the typical velocity d/τ must be much smaller than the speed of light; i.e., $d \ll c/\omega$, where ω is the angular frequency of the oscillation. Putting all these together, we arrive at the following conditions:

$$d \ll c/\omega \ll r \ . \quad (7.5)$$

When we are sufficiently far away from the sources as given by Eq. (7.5), the conservation of energy requires that the total power passing through any spherical surface enclosing the charges must be the same. Since the surface area for a sphere is proportional to r^2, if the Poynting vector drops faster than $1/r^2$, the total power drops to zero for $r \to \infty$. In other words, any term that decreases faster than $1/r^2$ will be negligible, and will not contribute to radiation.

While the field from localized sources could be expanded in a series of multipole-terms, in this chapter we will only consider radiation from electric and magnetic dipole terms.

7.3.1 Radiation from an electric point dipole

We will now investigate the radiation generated by the electric dipole moment of a localized charge distribution. As a concrete example, we consider a dipole consisting of two point charges $q_\pm$, situated on the z-axis at points $\pm\frac{1}{2}\,\hat{z}\,d$, as shown in Fig. 7.1. These charges are connected by a conducting wire. We assume that they oscillate sinusoidally:

$$q_\pm = \pm\, q_o \cos\omega t \ .$$

This time-dependence of charges $q_\pm$ is due to a current $\vec{\imath}$ that continuously redistributes the charges at the two end points thorough the connecting wire. This current is defined by

$$\vec{\imath} = \frac{dq_+}{dt}\,\hat{z} = -\,q_o\,\omega\,\sin\omega t\;\hat{z} \ .$$

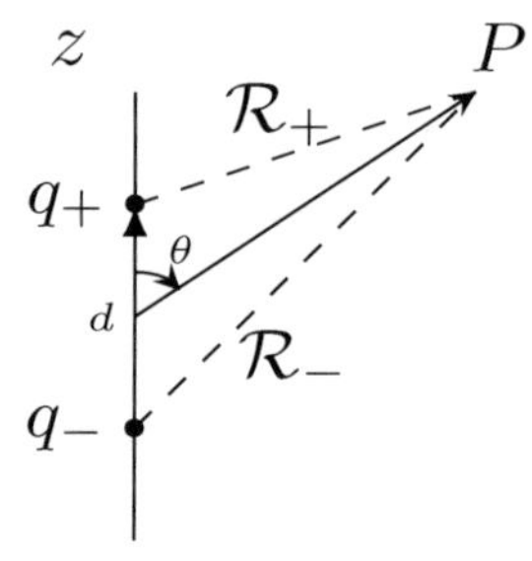

FIGURE 7.1
Two time varying charges $q_\pm = \pm q_o \cos\omega t$, connected by a wire, situated along the z-axis at $(0, 0, \pm\, d/2)$. We look for the potentials at point P situated at $\vec{r}$, where $|\vec{r}\,| \gg c/\omega \gg d$.

In the limit $d \to 0$, while the product $q_o d$ remains fixed, this system is called a *Hertzian dipole.*

From Eq. (7.4), the electric potential at point P due to this dipole is then given by

$$\begin{aligned} V(\vec{r},t) &= \frac{1}{4\pi\epsilon_o}\int_{\mathcal{V}} \frac{\rho(\vec{r}\,',t_r)}{|\vec{r}-\vec{r}\,'|}\, d^3r' \\ &= \frac{q_o}{4\pi\epsilon_o}\left[\frac{\cos\omega\left(t-\frac{|\vec{r}-\hat{z}\,d/2|}{c}\right)}{|\vec{r}-\hat{z}\,d/2|} - \frac{\cos\omega\left(t-\frac{|\vec{r}+\hat{z}\,d/2|}{c}\right)}{|\vec{r}+\hat{z}\,d/2|}\right], \end{aligned} \tag{7.6}$$

where the distances $\mathcal{R}_\pm$ of charges $q_\pm$ from the field point P are

$$\mathcal{R}_\pm \equiv |\vec{r}\mp\frac{1}{2}\,\hat{z}\,d| \approx \sqrt{r^2\mp d\,(\vec{r}\cdot\hat{z})} \approx r\left(1\mp\frac{1}{2}\frac{d}{r}\,(\hat{r}\cdot\hat{z})\right) = r\left(1\mp\frac{1}{2}\,\frac{d\cos\theta}{r}\right).$$

> **Exercise 7.1** Expressing the charge density of the configuration shown in Fig. 7.1 as a sum of δ-functions, prove Eq. (7.6).

Then, the approximate values for the terms in the square bracket in Eq. (7.6) can be written as

$$\begin{aligned} &\frac{\cos\omega\left(t-\frac{|\vec{r}\mp\frac{1}{2}\,\hat{z}\,d|}{c}\right)}{|\vec{r}\mp\frac{1}{2}\,\hat{z}\,d|} \approx \frac{\cos\omega\left(t-\frac{r}{c}\right)\mp\frac{1}{2}\left(\frac{\omega d\cos\theta}{c}\right)\sin\omega\left(t-\frac{r}{c}\right)}{r\left(1\mp\frac{1}{2}\,\frac{d\cos\theta}{r}\right)} \\ &\approx \frac{1}{r}\left[\cos\omega\left(t-\frac{r}{c}\right)\mp\frac{\omega d\cos\theta}{2c}\sin\omega\left(t-\frac{r}{c}\right)\right]\left(1\pm\frac{d\cos\theta}{2r}\right) \\ &\approx \frac{1}{r}\left[\cos\omega\left(t-\frac{r}{c}\right)\pm\frac{d\cos\theta}{2r}\,\cos\omega\left(t-\frac{r}{c}\right)\mp\frac{\omega d\cos\theta}{2c}\sin\omega\left(t-\frac{r}{c}\right)\right], \end{aligned}$$

where we have used $d \ll c/\omega \ll r$. Substituting these in Eq. (7.6), gives

$$\begin{aligned} V(\vec{r},t) &= \frac{q_o}{4\pi\epsilon_o}\,\frac{d\cos\theta}{r}\left[\frac{1}{r}\,\cos\omega\left(t-\frac{r}{c}\right)-\frac{\omega}{c}\sin\omega\left(t-\frac{r}{c}\right)\right] \\ &= -\frac{q_o}{4\pi\epsilon_o}\,\frac{d\cos\theta}{r}\,\frac{\omega}{c}\sin\left[\omega\left(t-\frac{r}{c}\right)\right]. \end{aligned} \tag{7.7}$$

Next, we derive the vector potential generated by the current flowing between the two charges. We start with Eq. (7.4), with $\vec{r}-\vec{r}\,'=\vec{r}-z'\hat{z}$, and hence $|\vec{r} - \vec{r}\,'| \approx r\,(1 - z'\cos\theta/r)$. Thus

$$\begin{aligned}
\vec{A}(\vec{r},t) &= \frac{\mu_o}{4\pi}\int_{\mathcal{V}} \frac{\vec{J}(\vec{r}\,',t_r)}{|\vec{r}-\vec{r}\,'|}\,d^3r' \approx \frac{\mu_o}{4\pi}\,\frac{\hat{z}}{r}\int_{-d/2}^{d/2} \frac{i(z',t_r)}{1 - z'/r\,\cos\theta}\,dz' \\
&\approx \frac{\mu_o}{4\pi}\,\frac{\hat{z}}{r}\int_{-d/2}^{d/2} \frac{-\,q_o\,\omega\,\sin\left[\omega\left(t - \frac{r - z'\cos\theta}{c}\right)\right]}{1 - z'/r\,\cos\theta}\,dz' \\
&\approx -\frac{\mu_o}{4\pi}\,\frac{q_o\,\omega\,\hat{z}}{r}\int_{-d/2}^{d/2} \left[\sin\omega\left(t - \frac{r}{c}\right)\right. \\
&\qquad\left. + \left(\frac{\omega z'\cos\theta}{c}\right)\cos\omega\left(t - \frac{r}{c}\right)\right](1 + z'\cos\theta/r)\;dz' \\
&\approx -\frac{\mu_o}{4\pi}\,\frac{q_o\,\omega\,\hat{z}}{r}\int_{-d/2}^{d/2} \sin\left[\omega\left(t - \frac{r}{c}\right)\right]\;dz' \\
&\approx -\frac{\mu_o}{4\pi}\,\frac{q_o\,d\,\omega\,\hat{z}}{r}\sin\left[\omega\left(t - \frac{r}{c}\right)\right]\;, \qquad (7.8)
\end{aligned}$$

where we have dropped the terms that are suppressed by factors of $\omega d/c$ or d/r.

> **Exercise 7.2** Show that for a current i along a very thin wire, we can write $\vec{J}(\vec{r}\,',t_r)\,d^3r'$ as $i(z',t_r)\,dz'\,\hat{z}$

Thus the vector potential due to the electric dipole of two oscillating charges separated by the distance d, is

$$\vec{A}(\vec{r},t) = -\frac{\mu_o}{4\pi}\,\frac{q_o\,d\,\omega}{r}\left[\sin\omega\left(t - \frac{r}{c}\right)\right]\,\hat{z}\;. \qquad (7.9)$$

In spherical coordinates, $\hat{z} = \cos\theta\,\hat{r} - \sin\theta\,\hat{\theta}$, and hence the vector potential can be written as $\vec{A}(\vec{r},t) = A_r\hat{r} + A_\theta\,\hat{\theta}$, where

$$\begin{aligned}
A_r &= -\frac{\mu_o}{4\pi}\,\frac{q_o\,d\,\omega}{r}\left[\sin\omega\left(t - \frac{r}{c}\right)\right]\cos\theta\;, \\
A_\theta &= \frac{\mu_o}{4\pi}\,\frac{q_o\,d\,\omega}{r}\left[\sin\omega\left(t - \frac{r}{c}\right)\right]\sin\theta\;.
\end{aligned}$$

The electric and magnetic fields at point $\vec{r}$ are given by

$$\vec{E} = -\vec{\nabla}\,V(\vec{r},t) - \frac{\partial}{\partial t}\vec{A}(\vec{r},t) \qquad \text{and} \qquad \vec{B} = \vec{\nabla}\times\vec{A}(\vec{r},t)\;,$$

which now reduce to

$$\vec{E} = -\frac{\mu_o\,q_o\,\omega^2 d}{4\pi r}\sin\theta\cos\left[\omega\left(t - \frac{r}{c}\right)\right]\;\hat{\theta}$$

and

$$\vec{B} = -\frac{\mu_o\, q_o\, \omega^2 d}{4\pi c r} \sin\theta \cos\left[\omega\left(t - \frac{r}{c}\right)\right]\, \hat{\phi}$$

where in deriving $\vec{E}$ we have used $c^2 \mu_o \epsilon_o = 1$.

Having calculated the fields, we now turn our attention to the power radiated by the dipole. The Poynting vector, $\vec{S} = \frac{1}{\mu_o}\vec{E} \times \vec{B}$, becomes

$$\vec{S} = \frac{\mu_o}{c} \frac{\left(q_o\, \omega^2 d\right)^2}{16\pi^2\, r^2} \sin^2\theta \cos^2\left[\omega\left(t - \frac{r}{c}\right)\right]\, \hat{r}\ . \tag{7.10}$$

Since $\langle \cos^2 x \rangle = 1/2$ where $\langle \cdot \rangle$ denotes the average over a period, the time average of the Poynting vector is

$$\langle \vec{S} \rangle = \frac{\mu_o}{c} \frac{\left(q_o\, \omega^2 d\right)^2}{32\pi^2} \frac{\sin^2\theta}{r^2}\, \hat{r}\ . \tag{7.11}$$

Fig. 7.2 shows the angular distribution of the radiated energy. It illustrates an interesting property of the oscillating dipole, i.e., it does not radiate energy along the direction of the acceleration, which in our case is $\hat{z}$.

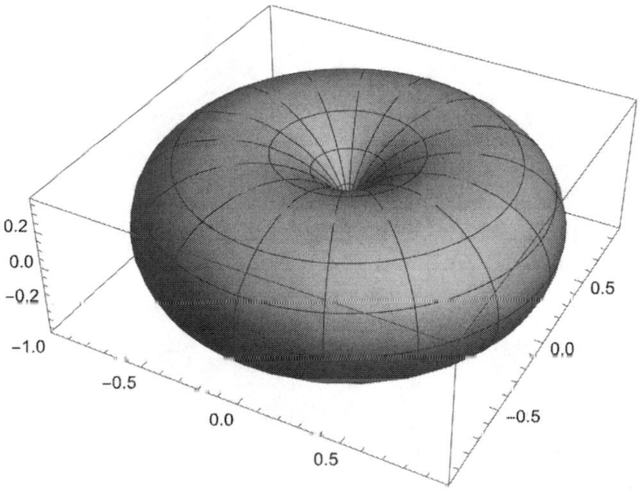

FIGURE 7.2
The angular distribution of the radiated energy about the axis of symmetry of the dipole. The power/area for a given radius is zero along the axis of the dipole and maximum on the $\theta = \pi/2$ plane.

The total power radiated through a large spherical surface is then given by

$$\mathscr{P} = \frac{\mu_o}{c} \frac{\left(q_o\, \omega^2 d\right)^2}{32\pi^2} \oint \frac{\sin^2\theta}{r^2}\, r^2 \sin\theta\, d\theta\, d\phi = \frac{\mu_o \left(q_o\, \omega^2 d\right)^2}{12\pi\, c}\ . \tag{7.12}$$

Since the charge flowing in the wire has an acceleration of the order of $\omega^2 d$, we see that the power radiated is proportional to the square of the acceleration.

7.3.2 Radiation from magnetic dipole

In this section we will investigate the radiation generated by a magnetic dipole. As a concrete example, we consider a square loop of dimension b lying on the xy-plane with sides parallel to the Cartesian axes as shown in Fig. 7.3. A current $i(t) = I_o \cos\omega t$ flows through the sides of the loop.

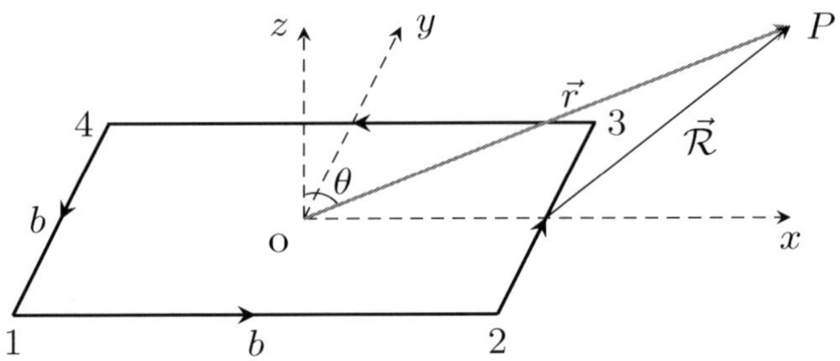

FIGURE 7.3
A square loop of side b carries a current $i = I_o \cos\omega t$. We are seeking the electromagnetic field at point $\vec{r}$.

Using Eq. (7.9), we can determine the vector potential for each side, and hence for the entire loop. The vector potential $\vec{A}_{23}(\vec{r}, t)$ due to side 2-3 is given by

$$\vec{A}_{23}(\vec{r}, t) = \frac{\mu_o}{4\pi} \frac{I_o\, b}{\mathcal{R}} \left[\cos\omega\left(t - \frac{\mathcal{R}}{c}\right)\right] \hat{y}\ ; \tag{7.13}$$

where $\mathcal{R} = \sqrt{\left(\vec{r} - \frac{b}{2}\hat{x}\right)^2} \approx r\sqrt{1 - b/r \sin\theta\cos\phi} \approx \left(r - \frac{1}{2}\, b\sin\theta\cos\phi\right)$, see Fig. 7.3. The factor $\frac{1}{\mathcal{R}}\left[\cos\omega\left(t - \frac{\mathcal{R}}{c}\right)\right]$ in Eq. (7.13) can then be expanded as

$$\begin{aligned}
\frac{1}{\mathcal{R}}\left[\cos\omega\left(t - \frac{\mathcal{R}}{c}\right)\right] &\approx \frac{1}{r}\left[\cos\omega\left(t - \frac{\left(r - \frac{1}{2}b\sin\theta\cos\phi\right)}{c}\right)\right] \\
&\quad\times\left(1 + \frac{b}{2r}\sin\theta\cos\phi\right) \\
&\approx \frac{1}{r}\left[\cos\omega\left(t - \frac{r}{c}\right) - \frac{\omega b\sin\theta\cos\phi}{2c}\sin\omega\left(t - \frac{r}{c}\right)\right] \\
&\quad\times\left(1 + \frac{b}{2r}\sin\theta\cos\phi\right) .
\end{aligned} \tag{7.14}$$

Exercise 7.3 Derive Eq. (7.14).

Thus, the vector potential due to the side 2-3 is

$$\vec{A}_{23}(\vec{r},t) \approx \frac{\mu_o}{4\pi}\frac{I_o\,b}{r}\left[\cos\omega\left(t-\frac{r}{c}\right)-\frac{\omega\,b\sin\theta\cos\phi}{2c}\sin\omega\left(t-\frac{r}{c}\right)\right] \times\left(1+\frac{b}{2r}\sin\theta\cos\phi\right)\,\hat{y}\,. \tag{7.15}$$

To determine the contribution of the segment 4-1, we use Eq. (7.15), where we replace b by $-b$. This replacement is necessary because the current in the segment 4-1 flows in a direction opposite to the current in segment 2-3, and the segment 4-1 is displaced by $-b\,\hat{x}$. Thus, we obtain

$$\vec{A}_{41}(\vec{r},t) \approx -\frac{\mu_o}{4\pi}\frac{I_o\,b}{r}\left[\cos\omega\left(t-\frac{r}{c}\right)+\frac{\omega b\sin\theta\cos\phi}{2c}\sin\omega\left(t-\frac{r}{c}\right)\right] \times\left(1-\frac{b}{2r}\sin\theta\cos\phi\right)\,\hat{y}\,. \tag{7.16}$$

Adding $\vec{A}_{23}(\vec{r},t)$ and $\vec{A}_{41}(\vec{r},t)$ and incorporating the approximations given in Eq. (7.5), the net contribution of the two sides that are parallel to the y-axis is

$$\vec{A}(\vec{r},t)_y = -\frac{\mu_o}{4\pi}\frac{I_o\,b}{r}\frac{\omega b\sin\theta\cos\phi}{2c}\sin\omega\left(t-\frac{r}{c}\right)\,\hat{y}\,. \tag{7.17}$$

Similarly, the contribution to the vector potential of the two sides parallel to x-axis is

$$\vec{A}(\vec{r},t)_x = \frac{\mu_o}{4\pi}\frac{I_o\,b}{r}\frac{\omega b\sin\theta\sin\phi}{2c}\sin\omega\left(t-\frac{r}{c}\right)\,\hat{x}\,. \tag{7.18}$$

Finally, putting the two contributions together, we obtain the net vector potential, $\vec{A}(\vec{r},t)=\vec{A}(\vec{r},t)_x+\vec{A}(\vec{r},t)_y$ due to the entire loop:

$$\begin{aligned}\vec{A}(\vec{r},t) &= \frac{\mu_o}{4\pi}\frac{I_o\,b^2\,\omega}{c}\frac{\sin\theta}{r}\sin\omega\left(t-\frac{r}{c}\right)\,(\sin\phi\hat{x}-\cos\phi\hat{y})\\ &= -\frac{\mu_o}{4\pi}\frac{m_o\omega}{c}\frac{\sin\theta}{r}\sin\omega\left(t-\frac{r}{c}\right)\,\hat{\phi}\,,\end{aligned} \tag{7.19}$$

where we have replaced $I_o b^2$ by m_o, the magnitude of the magnetic dipole moment of the loop.

Note that the contribution of the two sides parallel to y-axis given by Eq. (7.17), is smaller than magnitude of the vector potentials $\vec{A}_{23}(\vec{r},t)$ or $\vec{A}_{41}(\vec{r},t)$ generated by individual sides, or equivalently for an electric dipole, Eq. (7.9), by a factor $\omega b/c$. Hence, the contribution due to a magnetic dipole is much smaller than an electric dipole of comparable dimension.

We now compute the $\vec{E}$ and $\vec{B}$ fields from the vector potential above. For $\vec{E}$, the vector potential suffices because current is assumed to be the same everywhere within the loop, and hence there is no accumulation of charge anywhere. Hence, in accordance with Eq. (7.5), we have

$$\vec{E} = -\frac{\partial \vec{A}}{\partial t} = \frac{\mu_o m_o \omega^2}{4\pi c} \frac{\sin\theta}{r} \cos\omega\left(t - \frac{r}{c}\right)\hat{\phi}$$

and

$$\vec{B} = \vec{\nabla} \times \vec{A}(\vec{r}, t) = -\frac{\mu_o m_o \omega^2}{4\pi c^2} \frac{\sin\theta}{r} \cos\omega\left(t - \frac{r}{c}\right)\hat{\theta}\,.$$

The Poynting vector due to the magnetic dipole is then given by

$$\vec{S} = \frac{1}{\mu_o}\vec{E} \times \vec{B} = \frac{\mu_o}{c^3}\left(\frac{m_o\omega^2}{4\pi}\right)^2 \frac{\sin^2\theta}{r^2} \cos^2\omega\left(t - \frac{r}{c}\right)\hat{r}\,. \tag{7.20}$$

The total power radiated (time-averaged) by the magnetic dipole is then given by

$$\begin{aligned} \mathscr{P} &= \frac{\mu_o}{2c^3}\left(\frac{m_o\omega^2}{4\pi}\right)^2 \oint \frac{\sin^2\theta}{r^2}\, r^2 \sin\theta\, d\theta\, d\phi \\ &= \frac{\mu_o\, m_o^2\, \omega^4}{12\pi c^3}. \end{aligned} \tag{7.21}$$

For magnetic and electric dipoles of similar dimensions, the ratio of the power radiated is

$$\frac{\mathscr{P}_{\text{m.d.}}}{\mathscr{P}_{\text{e.d.}}} = \frac{b^2\,\omega^2}{c^2} \ll 1\,.$$

Thus, the magnetic dipole radiation becomes significant only in the absence of electric dipoles. The magnetic dipole radiation plays an important role for astrophysical systems such as neutron stars.

Exercise 7.4 Prove

$$\frac{\mathscr{P}_{\text{m.d.}}}{\mathscr{P}_{\text{e.d.}}} = \frac{b^2\,\omega^2}{c^2} \ll 1\,.$$

7.4 Liénard–Wiechert potentials; fields due to moving point charges

Inside an X-Ray machine, electrons race down a tube and hit a target where they experience a tremendous deceleration. The result is a burst of X-rays coming out. This is how X-ray tubes work. In this section we will see how moving charges undergoing acceleration radiate energy.

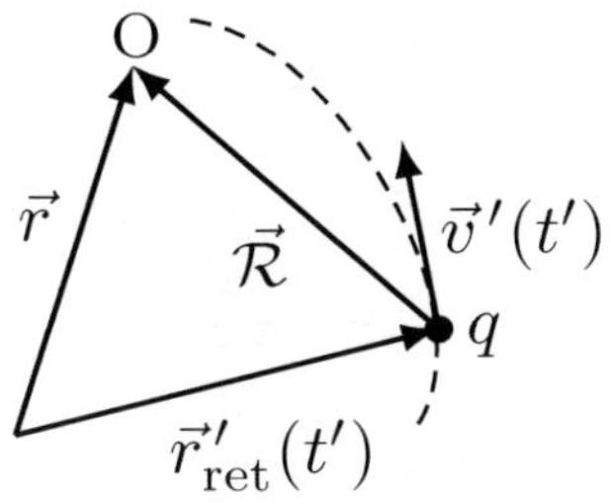

FIGURE 7.4
The radiation field due to a moving charge q is observed at time t at point O. The field at point O is due to the charge located at $\vec{r}\,'_{\rm ret}$, which the position the particle had at time $t' = t - \mathcal{R}/c$, where $\mathcal{R}$ is the magnitude of the vector $\vec{\mathcal{R}} = \vec{r} - \vec{r}\,'_{\rm ret}$. The dashed line shows the path of the particle.

We denote by $\vec{r}$ the point O where we are observing the radiation generated at time t due to a point charge q located at $\vec{r}\,'(t')$ and moving with velocity $\vec{v}\,'(t')$, as illustrated in Fig. 7.4.

From Eq. (7.4), we would be tempted to write the scalar potential at $\vec{r}$ at time t as

$$V(\vec{r}, t) = \frac{q}{4\pi\epsilon_{\rm o}\,|\vec{r} - \vec{r}\,'_{\rm ret}|}\,, \tag{7.22}$$

where $\vec{r}\,'_{\rm ret}$ denotes the position of the point charge at time $t - \mathcal{R}/c$ and $\vec{\mathcal{R}} = \vec{r} - \vec{r}\,'_{\rm ret}$. However, such an answer would be incorrect.

The above expression would have been correct if the spatial integral for scalar potential in Eq. (7.4) were to be carried out at the same instant of time. Instead, different parts of the space contributing to $V(\vec{r}, t)$ originate at different times. To derive the scalar potential at time t at the observation point O, we will need to include contributions from the entire space originating from non-zero charge density at different times. The contribution from a point at distance $\mathcal{R}$ from the observation point O, will be due to the source density at that point at an earlier time $t' = t - \mathcal{R}/c$.

In order to properly account for contribution from various parts of the space with source densities at an appropriate time, we draw very thin spherical shells around O of varying radii, each of infinitesimal thickness $c\,\delta t$. The farther a shell is from O, the earlier must the contribution originate from it to arrive at time t. [1]

It is important to note that we must visualize the particle as an extended object. The size of the particle plays no role in the final result, as we will see shortly. As the particle passes through these various shells, its contribution

[1] The graphical method we will use appears in several books including [14, 46].

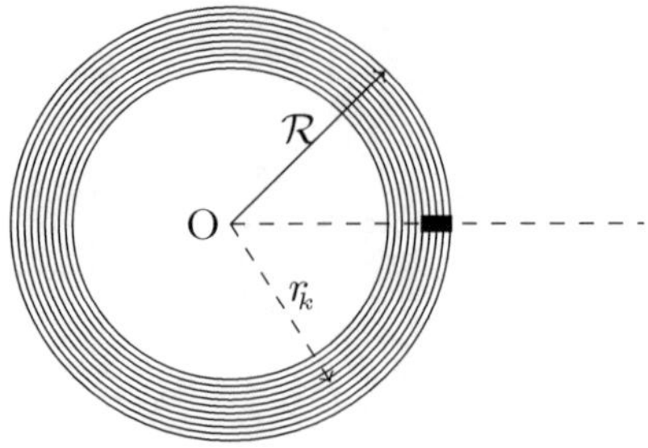

FIGURE 7.5
The charge is represented by the rectangle whose far end is at a distance $\mathcal{R}$ from the observation point O. The concentric shells have radii $r_k = \mathcal{R} - k\,c\,\delta t$, where δt is an infinitesimal time element and $k = 0, 1, 2, \ldots$. The velocity of the charge is to the left, towards O.

for each shell will be captured and added to the field at O to generate the field at time t.

While we aim to derive the expression for the scalar potential at O due a particle moving with an arbitrary velocity $\vec{v}\,'$ and acceleration $\vec{a}\,'$, we start with a special case in which $\vec{v}\,'$ is directed toward the observation point O.

We begin by assuming that the moving particle has a finite size, which for simplicity we show as a rectangular prism in Fig. 7.5. The concentric shells of radii $r_k = \mathcal{R} - k\,c\,\delta t$, where $k = 0, 1, 2, \ldots$, are the shells from where light takes $\mathcal{R}/c - k\,\delta t$ time to arrive at the observation point O at time t.

We assume that the area of cross-section of the prism is a^2 and the its length covers n-shells, and hence the length of the prism is $\ell = nc\,\delta t$, its volume ℓa^2 and its charge $q = \rho\,\ell a^2$.

In Fig. 7.5, the last shell of radius $\mathcal{R}$ lines up with the farthest end of the prism and the contribution to the potential at point O due to the charge within this shell takes a time $\mathcal{R}/c$ to reach O at time t. The light from the next inner shell takes δt less time to reach O, and hence the contribution to the potential from this shell must originate at time δt later to arrive at O at time t. During this time interval the charge moves a distance $v'\delta t$ closer to O, where v' is the speed of the particle directed towards the observation point. By the time the contribution from the back of the charge arrives to the front of the charge, the charge itself would have moved forward through Δn shells, where $\Delta n\, c\delta t = v'\,\frac{\ell}{c-v'}$. The net contribution to the potential $V(\mathrm{O}, t)$ at point O at time t is then given by

$$V(\mathrm{O}, t) = \frac{1}{4\pi\epsilon_o} \sum_{k=0}^{k=n+\Delta n} \frac{\rho a^2\, c\delta t}{r_k} \,.$$

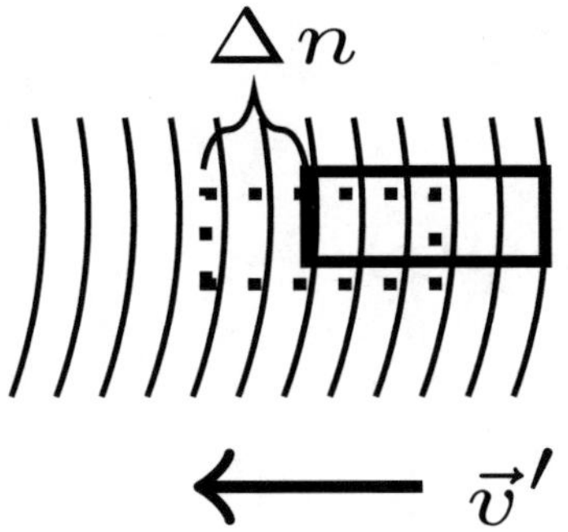

FIGURE 7.6
The darker rectangle shows the position of the charged particle when its rear end is at a distance $\mathcal{R}$ from the observation point O, which is depicted by a solid rectangle in Fig. 7.5. The dashed rectangle shows the position of the particle when the field originating from the rear end reaches the front end, and during this time the particle advances through Δn shells. The arrow shows the direction of the particle's velocity, which is directed towards point O.

For a point-like charge, we can assume that all $n + \Delta n$ shells have the same radius $\mathcal{R}$. This yields

$$\begin{aligned} V(\mathrm{O}, t) &= \frac{1}{4\pi\epsilon_o} \frac{(n + \Delta n) c \delta t \, \rho a^2}{\mathcal{R}} \\ &= \frac{1}{4\pi\epsilon_o} \frac{(\ell + v' \frac{\ell}{c - v'}) \rho a^2}{\mathcal{R}} \\ &= \frac{\rho \ell a^2}{4\pi\epsilon_o \mathcal{R}} \frac{1}{1 - v'/c} \\ &= \frac{q}{4\pi\epsilon_o \mathcal{R}} \frac{1}{1 - v'/c} \, . \end{aligned} \tag{7.23}$$

If the particle is not moving directly toward the observer, then the component of the velocity along the direction of the observer replaces v'. Hence, for a general velocity $\vec{v}'(t')$, the potential at point $\vec{r}$ at time t is given by

$$V(\vec{r}, t) = \frac{q}{4\pi\epsilon_o \mathcal{R}} \frac{1}{1 - \hat{\mathcal{R}} \cdot \vec{v}'/c} \equiv \frac{q}{4\pi\epsilon_o \mathcal{R}} \frac{1}{s} \, , \tag{7.24}$$

where $\vec{v}' = \partial \vec{r}'/\partial t'$ and we have defined $s = 1 - \hat{\mathcal{R}} \cdot \vec{v}'/c$. A similar reasoning leads to the following vector potential for the particle:

$$\vec{A}(\vec{r}, t) = \frac{\mu_o}{4\pi} \frac{q \vec{v}'}{\mathcal{R}} \frac{1}{1 - \hat{\mathcal{R}} \cdot \vec{v}'/c} \equiv \frac{\mu_o}{4\pi} \frac{q \vec{v}'}{\mathcal{R}} \frac{1}{s} \tag{7.25}$$

$$= \frac{\vec{v}'}{c^2} V(\vec{r}, t) \, . \tag{7.26}$$

This connection between $\vec{A}$ and V is not unexpected, because for a point-like charge, the source densities in Eq. (7.4) become δ-functions. These potentials due to a moving charge are known as the Liénard–Wiechert potentials.

7.4.1 Electromagnetic fields from Liénard–Wiechert potentials

Having derived the Liénard–Wiechert potentials we will now determine the electromagnetic fields using

$$\vec{E}(\vec{r},t) = -\vec{\nabla} V(\vec{r},t) - \partial\vec{A}(\vec{r},t)/\partial t \qquad ; \qquad \vec{B}(\vec{r},t) = \vec{\nabla} \times \vec{A}(\vec{r},t) \ .$$

However, these derivatives are not easy to compute. The complications arise from the non-trivial dependence of $\vec{\mathcal{R}} \equiv \vec{r} - \vec{r}'_{\text{ret}}$ on variables $\vec{r}$ and t, where $\vec{r}$ is the observation point and $\vec{r}'_{\text{ret}}$ denotes the position of the charge at time $t' = t - |\vec{r} - \vec{r}'_{\text{ret}}(t')|\,/c$. The variable $\vec{\mathcal{R}}$, and in particular the combination $s\mathcal{R}$ appears in the Liénard–Wiechert potentials given by Eqs. (7.24) and (7.25). To avoid getting lost in these differentiations and losing sight of our goal of deriving the electromagnetic fields from the accelerated charge, we will develop some of the necessary mathematical preliminaries in the next subsection before delving into the gory computation leading to $\vec{E}$ and $\vec{B}$ fields.

7.4.1.1 Mathematical preliminaries

In order to compute the electromagnetic fields, we have to calculate the time- and space-derivatives of $V(\vec{r},t)$ and $\vec{A}(\vec{r},t)$. For this, we first determine $\partial t'/\partial t$, where

$$t' = t - \frac{\mathcal{R}}{c} \ ,$$

and then the spatial derivative $\vec{\nabla} t'$. This will then be followed by calculating $\partial(s\mathcal{R})/\partial t'$, since $s\mathcal{R}$ is an important factor for the Liénard–Wiechert potentials (Eqs. 7.24 and 7.25).

For the time derivative we have

$$\frac{\partial t'}{\partial t} = 1 - \frac{1}{c}\frac{\partial \mathcal{R}}{\partial t} = 1 - \frac{1}{c}\frac{\partial}{\partial t'}\left|\vec{r} - \vec{r}'_{\text{ret}}(t')\right| \frac{\partial t'}{\partial t} \ .$$

Solving for $\partial t'/\partial t$, we obtain the time derivative

$$\frac{\partial t'}{\partial t} = \frac{1}{1 - \hat{\mathcal{R}}\cdot\vec{v}'/c} = \frac{1}{s} \ . \tag{7.27}$$

In our derivation we have used

$$\frac{\partial}{\partial t'}\left|\vec{r} - \vec{r}'_{\text{ret}}(t')\right| = \frac{\partial \mathcal{R}}{\partial t'} = -\hat{\mathcal{R}}\cdot\vec{v}' \ .$$

Exercise 7.5 Prove that

$$\frac{\partial \mathcal{R}}{\partial t'} = -\hat{\mathcal{R}} \cdot \vec{v}' \,.$$

We next determine the spatial derivative $\vec{\nabla} t'$ using

$$\vec{\nabla} t' = \vec{\nabla}\left(t - \mathcal{R}/c\right) = -\frac{1}{c}\vec{\nabla}\mathcal{R} = -\frac{\hat{\mathcal{R}}}{c} - \frac{1}{c}\left(-\hat{\mathcal{R}} \cdot \vec{v}'\right)\vec{\nabla} t' \,,$$

where we have used $\vec{\nabla}\mathcal{R} = \hat{\mathcal{R}} + \frac{\partial \mathcal{R}}{\partial t'}\vec{\nabla} t'$. Solving the above equation for $\vec{\nabla} t'$, we get the space derivative

$$\vec{\nabla} t' = \frac{-\hat{\mathcal{R}}}{c\left(1 - \hat{\mathcal{R}} \cdot \vec{v}'/c\right)} = -\frac{\hat{\mathcal{R}}}{c\,s}. \tag{7.28}$$

Exercise 7.6 Explicitly computing the x-component $\frac{\partial \mathcal{R}}{\partial x}$, show that the gradient of $\mathcal{R}$ is given by

$$\vec{\nabla}\mathcal{R} = \hat{\mathcal{R}} + \frac{\partial \mathcal{R}}{\partial t'}\vec{\nabla} t' \,.$$

Note that this result could also be derived using

$$\vec{\nabla}\mathcal{R} = \left.\vec{\nabla}\mathcal{R}\right|_{t'=\text{constant}} + \frac{\partial \mathcal{R}}{\partial t'}\vec{\nabla} t' \,.$$

Exercise 7.7 Show that

$$\hat{\mathcal{R}} \times \left(\hat{\mathcal{R}} \times \vec{v}'\right) = \left(\vec{v}' \cdot \hat{\mathcal{R}}\right)\hat{\mathcal{R}} - \vec{v}' \,,$$

and hence it follows that

$$\vec{v}' \cdot \left[\hat{\mathcal{R}} \times \left(\hat{\mathcal{R}} \times \vec{v}'\right)\right] = \left(\vec{v}' \cdot \hat{\mathcal{R}}\right)^2 - (v')^2 \,.$$

Exercise 7.8 Show that

$$\frac{\partial \hat{\mathcal{R}}}{\partial t'} = \frac{\hat{\mathcal{R}} \times \left(\hat{\mathcal{R}} \times \vec{v}'\right)}{\mathcal{R}} \,.$$

Now we are ready to differentiate $\mathcal{R}\left(1-\hat{\mathcal{R}}\cdot\vec{v}\,'/c\right)=\mathcal{R}\,s$.

$$\begin{aligned}
\frac{\partial}{\partial t'}\left[\mathcal{R}\left(1-\frac{\hat{\mathcal{R}}\cdot\vec{v}\,'}{c}\right)\right] &= \frac{\partial\mathcal{R}}{\partial t'}\,s+\mathcal{R}\,\frac{\partial s}{\partial t'}\\
&= -\hat{\mathcal{R}}\cdot\vec{v}\,'\left(1-\frac{\hat{\mathcal{R}}\cdot\vec{v}\,'}{c}\right)-\mathcal{R}\left(\frac{\hat{\mathcal{R}}\cdot\vec{a}\,'}{c}+\frac{\vec{v}\,'}{c}\cdot\frac{\hat{\mathcal{R}}\times\left(\hat{\mathcal{R}}\times\vec{v}\,'\right)}{\mathcal{R}}\right)\\
&= -\hat{\mathcal{R}}\cdot\vec{v}\,'\left(1-\frac{\hat{\mathcal{R}}\cdot\vec{v}\,'}{c}\right)-\frac{\vec{\mathcal{R}}\cdot\vec{a}\,'}{c}-\frac{\vec{v}\,'}{c}\cdot\left(\hat{\mathcal{R}}\times\left(\hat{\mathcal{R}}\times\vec{v}\,'\right)\right)\\
&= -\frac{\vec{\mathcal{R}}\cdot\vec{a}\,'}{c}+\frac{\vec{v}\,'\cdot\vec{v}\,'}{c}-\hat{\mathcal{R}}\cdot\vec{v}\,'\,,
\end{aligned}$$

where we have used $\frac{\partial s}{\partial t'}=-\left(\frac{\hat{\mathcal{R}}\cdot\vec{a}\,'}{c}+\frac{\vec{v}\,'}{c}\cdot\frac{\hat{\mathcal{R}}\times\left(\hat{\mathcal{R}}\times\vec{v}\,'\right)}{\mathcal{R}}\right)$ and $\vec{a}\,'=\frac{\partial\vec{v}\,'}{\partial t'}$. We finally obtain

$$\frac{\partial}{\partial t'}\left(s\mathcal{R}\right)=-\frac{\vec{\mathcal{R}}\cdot\vec{a}\,'}{c}-\vec{v}\,'\cdot\left(\hat{\mathcal{R}}-\vec{\beta}'\right)\,, \tag{7.29}$$

where $\vec{\beta}'\equiv\frac{\vec{v}\,'}{c}$.

Exercise 7.9 Prove that

$$\frac{\partial s}{\partial t'}=-\left(\frac{\hat{\mathcal{R}}\cdot\vec{a}\,'}{c}+\vec{\beta}'\cdot\frac{\hat{\mathcal{R}}\times\left(\hat{\mathcal{R}}\times\vec{v}\,'\right)}{\mathcal{R}}\right)$$

7.4.1.2 The radiation fields

With mathematical preliminaries covered, we are now able to determine the $\vec{E}$ and $\vec{B}$ fields from the Liénard–Wiechert potentials.

The electric field $\vec{E}$ radiated by a moving charge

The electric field is given by $\vec{E}=-\vec{\nabla}V(\vec{r},t)-\partial\vec{A}(\vec{r},t)/\partial t$. Hence we first compute the gradient of the scalar field $V(\vec{r},t)$, which is given by

$$\vec{\nabla}V=\vec{\nabla}V\Big|_{t'=\text{constant}}+\frac{\partial V}{\partial t'}\,\vec{\nabla}t'\,. \tag{7.30}$$

Exercise 7.10 Explicitly computing $\frac{\partial V}{\partial x}$, verify Eq. (7.30).

$$\vec{\nabla} V\Big|_{\text{const. } t'} = \vec{\nabla}\left(\frac{1}{4\pi\epsilon_o}\frac{q}{s\mathcal{R}}\right)\Big|_{\text{const. } t'} = -\frac{q}{4\pi\epsilon_o}\frac{\left(\hat{\mathcal{R}} - \vec{\beta}'\right)}{\mathcal{R}^2\, s^2}$$

$$\begin{aligned}
\frac{\partial V}{\partial t'} &= -\frac{q}{4\pi\epsilon_o}\frac{1}{\mathcal{R}^2\, s^2}\frac{\partial}{\partial t'}(s\mathcal{R}) \\
&= -\frac{q}{4\pi\epsilon_o}\frac{1}{\mathcal{R}^2\, s^2}\left[-\mathcal{R}\frac{\hat{\mathcal{R}}\cdot\vec{a}'}{c} - \vec{v}'\cdot\left(\hat{\mathcal{R}} - \vec{\beta}'\right)\right]
\end{aligned}$$

Combining these and substituting for $\vec{\nabla} t'$, we get

$$\begin{aligned}
\vec{\nabla} V &= -\frac{q}{4\pi\epsilon_o}\frac{\left(\hat{\mathcal{R}} - \vec{\beta}'\right)}{\mathcal{R}^2\, s^2} \\
&\quad + \left(\frac{-\hat{\mathcal{R}}}{c\, s}\right)\left(-\frac{q}{4\pi\epsilon_o}\frac{1}{\mathcal{R}^2\, s^2}\left[-\mathcal{R}\frac{\hat{\mathcal{R}}\cdot\vec{a}'}{c} - \vec{v}'\cdot\left(\hat{\mathcal{R}} - \vec{\beta}'\right)\right]\right) \\
&= -\frac{q}{4\pi\epsilon_o}\frac{1}{s^3\mathcal{R}^2}\left[\left(1 - \hat{\mathcal{R}}\cdot\vec{\beta}'\right)\left(\hat{\mathcal{R}} - \vec{\beta}'\right) - \vec{v}'\cdot\left(\hat{\mathcal{R}} - \vec{\beta}'\right)\left(-\frac{\hat{\mathcal{R}}}{c}\right)\right] \\
&\quad - \frac{q}{4\pi\epsilon_o}\frac{1}{s^3\mathcal{R}^2}\left(-\mathcal{R}\frac{\hat{\mathcal{R}}\cdot\vec{a}'}{c}\right)\left(\frac{-\hat{\mathcal{R}}}{c}\right) \\
&= -\frac{q}{4\pi\epsilon_o\, s^3}\left(\frac{\hat{\mathcal{R}}\left(1 - \beta'^2\right) - \vec{\beta}'\left(1 - \hat{\mathcal{R}}\cdot\vec{\beta}'\right)}{\mathcal{R}^2} + \frac{\left(\hat{\mathcal{R}}\cdot\vec{a}'\right)\hat{\mathcal{R}}}{c^2\,\mathcal{R}}\right).
\end{aligned}$$

The other piece needed to compute the $\vec{E}$-field is $\partial\vec{A}(\vec{r},t)/\partial t$. It is given by

$$\begin{aligned}
\frac{\partial\vec{A}(\vec{r},t)}{\partial t} &= \frac{\partial\vec{A}}{\partial t'}\frac{\partial t'}{\partial t} = \frac{\partial}{\partial t'}\left(\frac{\mu_o}{4\pi}\frac{q\,\vec{v}'}{\mathcal{R}}\frac{1}{1 - \hat{\mathcal{R}}\cdot\vec{\beta}'}\right)\frac{1}{s} \\
&= \frac{\mu_o}{4\pi\, s^3}\left[\frac{q\,\vec{a}'}{\mathcal{R}}\left(1 - \hat{\mathcal{R}}\cdot\vec{\beta}'\right)\right. \\
&\quad \left. - \frac{q\,\vec{v}'}{\mathcal{R}^2}\left(-\mathcal{R}\frac{\hat{\mathcal{R}}\cdot\vec{a}'}{c} - \vec{v}'\cdot\left(\hat{\mathcal{R}} - \vec{\beta}'\right)\right)\right] \\
&= \frac{q}{4\pi\epsilon_o\, c^2 s^3}\left[\frac{\vec{a}'}{\mathcal{R}}\left(1 - \hat{\mathcal{R}}\cdot\vec{\beta}'\right) - \vec{v}'\left(-\frac{\hat{\mathcal{R}}\cdot\vec{a}'}{c\mathcal{R}} - \frac{\vec{v}'}{\mathcal{R}^2}\cdot\left(\hat{\mathcal{R}} - \vec{\beta}'\right)\right)\right],
\end{aligned}$$

where we replaced μ_o by $1/(4\pi\epsilon_o\, c^2)$. The electrical field is then given by

$$\vec{E} = \frac{q}{4\pi\epsilon_o\, s^3}\left[\frac{\hat{\mathcal{R}}\times\left(\vec{a}'\times\vec{\beta}'\right) - \left[\vec{a}' - \left(\vec{a}'\cdot\hat{\mathcal{R}}\right)\hat{\mathcal{R}}\right]}{c^2\mathcal{R}} + \frac{\left(\hat{\mathcal{R}} - \vec{\beta}'\right)\left(1 - \beta'^2\right)}{\mathcal{R}^2}\right]. \tag{7.31}$$

For $\beta' \to 0$ and $a' \to 0$, the electric field reduces to $q\hat{\mathcal{R}}/(4\pi\epsilon_o \mathcal{R}^2)$, as expected.

We can write the expression for the electric field as a sum of two parts

$$\vec{E} = \vec{E}_a + \vec{E}_v ,$$

where

$$\begin{aligned}
\vec{E}_a &= \frac{q}{4\pi\epsilon_o s^3} \frac{-\left[\vec{a}' - \left(\vec{a}' \cdot \hat{\mathcal{R}}\right)\hat{\mathcal{R}}\right] + \hat{\mathcal{R}} \times \left(\vec{a}' \times \vec{\beta}'\right)}{c^2 \mathcal{R}} ; \\
\vec{E}_v &= \frac{q}{4\pi\epsilon_o s^3} \frac{\left(\hat{\mathcal{R}} - \vec{\beta}'\right)\left(1 - \beta'^2\right)}{\mathcal{R}^2} .
\end{aligned} \tag{7.32}$$

We collected the terms that depend on acceleration in $\vec{E}_a$, and the remaining terms in $\vec{E}_v$. The major reason behind this separation, as we will soon see, is that $\vec{E}_v$ does not contribute to radiation, while $\vec{E}_a$ does. Note that $\vec{E}_a$ is perpendicular to $\hat{\mathcal{R}}$, an important characteristic of the radiation field.

Exercise 7.11 Show that the field $\vec{E}_a$ is perpendicular to $\hat{\mathcal{R}}$.

Exercise 7.12 Show that Eq. (7.31) under proper limits reduces to the appropriate expression for the electric field due to a stationary charge.

Exercise 7.13 The above expression for $\vec{E}$ can be written in an elegant and succinct form by defining a vector $\vec{u} = c\hat{\mathcal{R}} - \vec{v}'$. The quantity $\mathcal{R}\left(1 - \hat{\mathcal{R}} \cdot \vec{\beta}'\right)$ can then be written as

$$\mathcal{R}\left(1 - \hat{\mathcal{R}} \cdot \vec{\beta}'\right) = \frac{1}{c}\vec{\mathcal{R}} \cdot \vec{u} .$$

Show that using the vector $\vec{u}$, the $\vec{E}$-field can be written as [?]

$$\vec{E} = \frac{q}{4\pi\epsilon_o} \frac{\mathcal{R}}{(\vec{\mathcal{R}} \cdot \vec{u})^3} \left[\left(c^2 - v'^2\right)\vec{u} + \vec{\mathcal{R}} \times \left(\vec{u} \times \vec{a}'\right)\right] . \tag{7.33}$$

The magnetic field $\vec{B}$ radiated by a moving charge

The $\vec{B}$-field is similarly found by taking the curl of the vector potential

$$\vec{B} = \vec{\nabla} \times \vec{A}(\vec{r}, t) = \vec{\nabla} \times \vec{A}\Big|_{t'=\text{constant}} - \frac{\partial \vec{A}}{\partial t} \times \vec{\nabla} t' \,.$$

Exercise 7.14 Show that

$$\vec{\nabla} \times \vec{A}(\vec{r}, t) = \vec{\nabla} \times \vec{A}\Big|_{t'=\text{constant}} - \frac{\partial \vec{A}}{\partial t} \times \vec{\nabla} t' \,.$$

by explicitly computing the x-component.

A computation along the lines illustrated for $\vec{E}$ leads to

$$\vec{B} = \frac{q\,\mu_o}{4\pi} \left[\frac{\left(\vec{\beta}' \times \hat{\mathcal{R}}\right)\left(1 - \beta'^2\right)}{\mathcal{R}^2 s^3} + \hat{\mathcal{R}} \times \frac{\hat{\mathcal{R}} \times \left(\vec{a}' \times \vec{\beta}'\right) - \vec{a}'}{c^2 \mathcal{R} s^3} \right] . \tag{7.34}$$

As we did for the electric field, we separate $\vec{B}$ in terms that depend on acceleration and terms that do not: $\vec{B} = \vec{B}_a + \vec{B}_v$, where

$$\vec{B}_a = \frac{q\,\mu_o}{4\pi} \hat{\mathcal{R}} \times \frac{\hat{\mathcal{R}} \times \left(\vec{a}' \times \vec{\beta}'\right) - \vec{a}'}{c^2 \mathcal{R} s^3} \;;\; \vec{B}_v = \frac{q\,\mu_o}{4\pi} \frac{\left(\vec{\beta}' \times \hat{\mathcal{R}}\right)\left(1 - \beta'^2\right)}{\mathcal{R}^2 s^3} \,.$$

For the magnetic field, we see that both $\vec{B}_a$ and $\vec{B}_v$ are perpendicular to $\hat{\mathcal{R}}$. Furthermore, $\vec{B}$ vanishes as $\beta' \to 0$ and $a' \to 0$, indicating that a stationary charge does not generate a magnetic field.

Exercise 7.15 Show that the $\vec{B}$ and $\vec{E}$ fields from a moving charge are related by

$$\vec{B} = \frac{\hat{\mathcal{R}}}{c} \times \vec{E} \,.$$

We will next determine the radiated power.

7.4.1.3 The radiation from a non-relativistically moving charge

The power radiated by the particle is given by integrating the normal component of the Poynting vector $\vec{S} = \frac{1}{\mu_o} \vec{E} \times \vec{B}$ over a surface enclosing the charge, which for simplicity will be computed over the surface of a sphere of radius $\mathcal{R}$. Since $\vec{B} = \frac{1}{c} \hat{\mathcal{R}} \times \vec{E}$, the Poynting vector is given by

$$\vec{S} = \frac{1}{\mu_o c} \vec{E} \times \left(\hat{\mathcal{R}} \times \vec{E}\right) = \frac{1}{\mu_o c} \left[\hat{\mathcal{R}}\, E^2 - \left(\vec{E} \cdot \hat{\mathcal{R}}\right) \vec{E}\right] . \tag{7.35}$$

To go further with this computation, we note that fields $\vec{E}_a$ and $\vec{B}_a$ decrease as $1/\mathcal{R}$, while the acceleration independent fields $\vec{E}_v$ and $\vec{B}_v$ decrease as $1/\mathcal{R}^2$. Since the surface area is proportional to $\mathcal{R}^2$, to make a non-zero contribution to the power, the electric field should not decrease faster than $1/\mathcal{R}$. Thus, the terms that do not contain acceleration of the particle do not contribute to radiation. This is consistent with our earlier statements that only accelerated charges radiate. Hereafter, we drop all terms from $\vec{E}$ and $\vec{B}$ that do not contain acceleration. Thus, the Poynting vector reduces to

$$\vec{S} = \frac{\hat{\mathcal{R}}}{\mu_o c} \; E_a^2 \; . \tag{7.36}$$

At this point we assume that v', the speed of the particle, is very small as compared to speed of light c, so that $\beta' \ll 1$. With this approximation $s = \left(1 - \hat{\mathcal{R}} \cdot \vec{\beta}'\right) \to 1$, and the electric field

$$\vec{E}_a = \frac{q}{4\pi\epsilon_o \, s^3} \frac{-\left[\vec{a}' - \left(\vec{a}' \cdot \hat{\mathcal{R}}\right)\hat{\mathcal{R}}\right] + \hat{\mathcal{R}} \times \left(\vec{a}' \times \vec{\beta}'\right)}{c^2 \mathcal{R}}$$

reduces to

$$\vec{E}_a \simeq \frac{q}{4\pi\epsilon_o} \frac{-\left[\vec{a}' - \left(\vec{a}' \cdot \hat{\mathcal{R}}\right)\hat{\mathcal{R}}\right]}{c^2 \, \mathcal{R}} \; . \tag{7.37}$$

The vector $\vec{a}' - \left(\vec{a}' \cdot \hat{\mathcal{R}}\right)\hat{\mathcal{R}}$ is the projection of the acceleration on a plane perpendicular to $\hat{\mathcal{R}}$ and its magnitude is given by $a' \sin\theta$, where θ denotes the angle between vectors $\vec{a}'$ and $\hat{\mathcal{R}}$. As a result, the Poynting vector is given by

$$\vec{S} = \frac{\hat{\mathcal{R}}}{\mu_o c} \left[\frac{q}{4\pi\epsilon_o} \frac{a' \sin\theta}{c^2 \, \mathcal{R}}\right]^2 = \hat{\mathcal{R}} \, \frac{q^2}{16\pi^2\epsilon_o c^3} \frac{a'^2 \sin^2\theta}{\mathcal{R}^2} \; . \tag{7.38}$$

Eq. (7.38) shows that the power distribution about the direction of acceleration is of the same form as shown in Fig. 7.2. Consequently, there is no radiation along the direction of the acceleration, and the power distribution is maximum at an angle of $\pi/2$ from the acceleration.

Integrating the normal component of the Poynting vector over a sphere, we get

$$\oint \vec{S} \cdot \hat{\mathcal{R}} \, \mathcal{R}^2 d\Omega = \frac{q^2 \, a'^2}{16\pi^2\epsilon_o c^3} \oint \frac{\sin^2\theta}{\mathcal{R}^2} \mathcal{R}^2 \sin\theta d\theta d\phi = \frac{q^2 \, a'^2}{6\pi\epsilon_o c^3} \; .$$

This is known as the *Larmor formula* for radiation from a point charge.

7.5 Problems

7.1 The current on an infinite wire is switched on at time $t = 0$; i.e.,

$$I(t) = \begin{cases} 0 & t \leq 0 \\ I & t > 0 \end{cases}$$

Determine the potentials and electromagnetic field at a point at finite distance from the wire.

7.2 Why does sky appear blue while the Sun appears red during sunrise and sunset?

7.3 Derive the scalar potential given in Eq. (7.7) starting from the vector potential of Eq. (7.9) and the Lorentz condition $\vec{\nabla} \cdot \vec{A} + \mu_o \epsilon_o \frac{\partial V}{\partial t} = 0$.

7.4 Similar to the derivation given in Sec. (7.3.2), derive the vector potential due to a circular loop of radius R centered at the origin and lying on the xy-plane.

7.5 Expressing the total power radiated in terms of the current amplitude I_o, compare the total power radiated from an electric dipole with that from a magnetic dipole of similar dimension and frequency.

7.6 Denote the dipole moment of an electron in a hydrogen atom by

$$\vec{p} = q\,r \left[\hat{i} \cos \omega t + \hat{j} \sin \omega t\right] ,$$

where r is the classical Bohr radius.

(a) Determine the power radiated by the atom.

(b) Determine the life time of a classical hydrogen atom; i.e., the time it would take for the radius of the electronic orbit to shrink from $r_o = \frac{4\pi\epsilon_o\hbar^2}{m_e e^2}$ to zero.

8

Special Relativity

8.1 Introduction

Theory of relativity describes how measurements and observations can differ depending on the observer's reference frame. For example, a line charge at rest in a given frame of reference creates a static electric field. However, from the point of view of a moving frame relative to the first one, we would not only measure an electric field but also a magnetic field arising from the moving charge. Even if they look different, the two descriptions are essentially the same, as we will see shortly.

The theory of electromagnetism, embodied by Maxwell's equations, needs to be consistent across different inertial reference frames. Under Galilean relativity, Maxwell's equations do not retain their form, and hence, would not describe the same physics in both frames. This presented a dilemma: either modify electromagnetism or seek a new theory of relativity.

Faced with this challenge, Albert Einstein made a crucial decision. He opted to preserve the well-established Maxwell's equations, which implied the need for a new framework for understanding motion. This quest for consistency led to the development of *Special Relativity.*

This chapter delves into the fundamental principles of special relativity. We will derive the Lorentz transformations straight from Einstein's postulates. After deriving the Lorentz transformations we will cast Maxwell's equations in a covariant form, explicitly revealing their covariance. Moreover, we will illustrate how electromagnetic fields transform under change of reference frames.

8.2 Postulates of special relativity

8.2.1 Galilean relativity

Galileo's principle of relativity states that the laws of physics are the same in all inertial frames of reference. This means that if you are in a windowless room and cannot see outside, there is no experiment you can perform to determine if you are at rest or moving at a constant velocity.

DOI: 10.1201/9781003174967-8

Isaac Newton also made significant contributions to the concept of relativity. Newton's laws of motion together with the law of universal gravitation laid the groundwork for classical mechanics, which dominated physics for centuries. Within this framework, Newton introduced the concept that the laws of physics are the same for all observers in inertial frames of reference.

We denote by $\Delta\vec{r}$ the change in position of a particle during the time interval Δt as measured in the inertial frame S. Similarly, $\Delta\vec{r}'$ and $\Delta t'$, are the same quantities measured in the inertial frame S′, which is moving with velocity $\vec{v}_o$ relative to S. In Galilean relativity, these quantities are related by[1]

$$\Delta t' = \Delta t \; ; \quad \Delta\vec{r}' = \Delta\vec{r} - \vec{v}_o \Delta t \; . \tag{8.1}$$

These relations are known as Galilean transformations. The velocities $\vec{v}$ and $\vec{v}'$ of the particle are then related

$$\vec{v}' = \vec{v} - \vec{v}_o \; . \tag{8.2}$$

Thus, the speeds of the particle in the two inertial frames are related by

$$v' = |\vec{v} - \vec{v}_o| = \sqrt{v^2 + v_o^2 - 2\vec{v}\cdot\vec{v}_o} \; . \tag{8.3}$$

While this correlation between velocities aligned well with Newton's laws of motion, it was observed to have problem with electromagnetism.

Derived from Maxwell's equations, Eq. (4.13) indicates that electromagnetic waves, propagate at a speed $c = 1/\sqrt{\mu_o \epsilon_o}$. However, in the late 1880s, this generated a conundrum for physicists. Previously, all waves known to them needed a medium for propagation, raising the question: what medium carries light? This led to the postulation of a *Luminiferous aether* or simply ether, which was believed to carry the oscillations of the electromagnetic waves. According to this hypothesis, light waves traveled at the speed c in a reference frame at rest with respect to this all-pervading ether. However, in frames in motion relative to the ether, it was expected that the speed of light would vary depending on the relative velocity to the ether.

In 1887, two American physicists, Albert A. Michelson and Edward W. Morley, designed an ingenious experiment to detect any discrepancy in the speed of light when measured in directions aligned to, and perpendicular to Earth's motion through this ether. The experiment produced no observable variance. Michelson and Morley then repeated the experiment six months later to eliminate the possibility that the initial absence of any differential speed was merely coincidental with Earth being at rest relative to the ether at the time of the first experiment. Yet again, they failed to detect any variation in speed along directions parallel or perpendicular to Earth's velocity. Various explanations were proposed to account for this null outcome, but they became increasingly implausible, ultimately dealing a serious blow to ether theories.

[1]Note the linearity of these equations, which is crucial for maintaining the invertibility of the relationship.

Looking for simplicity, Einstein postulated that the speed of light was the same in all inertial frames. This, along with the principle of relativity and isotropy, led to the Lorentz transformations, which in turn left the form of the Maxwell's equations invariant. It could be argued that Einstein aligned himself more closely with Maxwell's theory of electromagnetism rather than the laws of classical mechanics, which coexisted harmoniously with Galilean relativity.

8.3 Einstein's postulates

Einstein postulated that:

1. The laws of physics are the same in all inertial frames.
2. The speed of light is the same in all inertial frames.

Since the second postulate contradicts Galilean relativity, which is grounded in our common day-to-day experience, it holds profound implications for space and time coordinates. We will now utilize these postulates to derive the *Lorentz transformations*, which describe the relationships between coordinates and times in two distinct inertial frames.

Let us consider two inertial frames S and S$'$ and assume that S$'$ moves with velocity $\vec{v}_o$ relative to S. For simplicity, and without loss of generality, we consider that at $t = t' = 0$ the axes of S and S$'$ coincide, and that S$'$ moves along the x-axis with speed v_o. See Fig. 8.1.

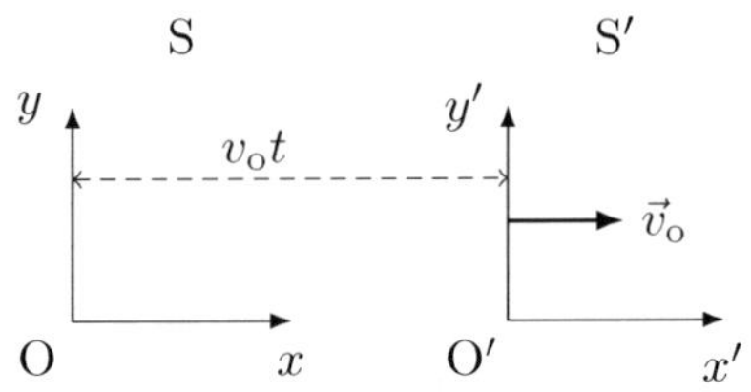

FIGURE 8.1
The inertial reference frame S$'$ has velocity $\vec{v}_o$ with respect to frame S, directed along the x-axis. At time t, the origin O$'$ is at a distance $v_o t$ from O.

8.3.1 Time dilation

We consider a light beam that travels a distance d' along y' axis in frame S$'$ where it is reflected and then travels back to its point of origin, as shown in

Fig. 8.2. Since the beam travels at speed c, the time $\Delta t'$ taken for this trip in frame S$'$ is $2d'/c$. When the same light beam is seen from frame S, the light

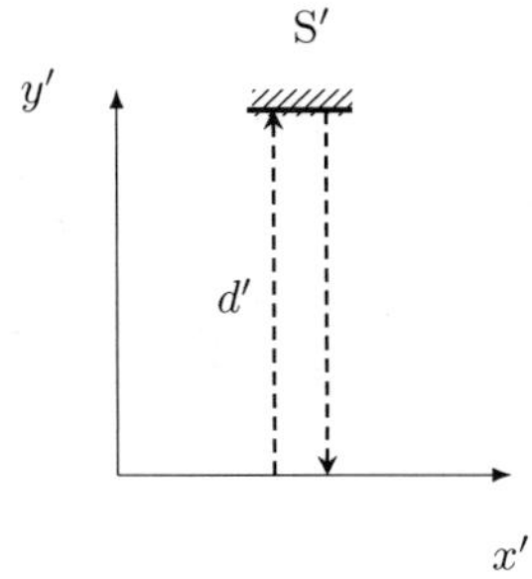

FIGURE 8.2
The diagram shows a light beam moving vertically up a distance d' in frame S$'$ and then reflected of a mirror. The light then travels back the same distance d' and is detected at the point of origin at time $\Delta t' = 2d'/c$.

no longer travels vertically. Instead, while the light moves toward the mirror, the mirror is moving along the positive x-axis with speed v_o, as shown in Fig. 8.3. If the time for the trip in frame S is Δt, the time for the light to reach

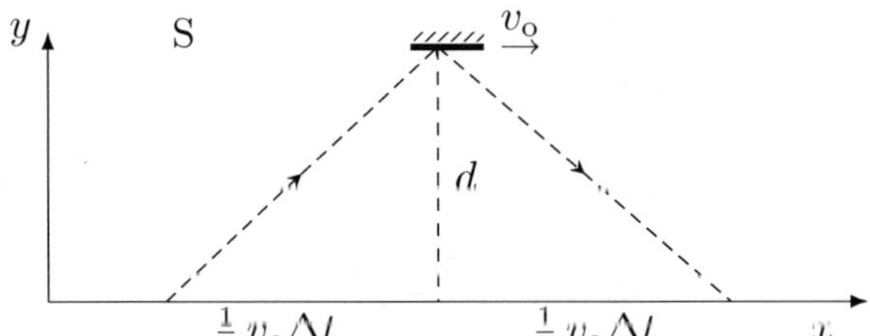

FIGURE 8.3
The diagram shows the path of the light beam as seen in frame S.

the mirror would be $\Delta t/2$. The total distance traveled by the light is then $2\sqrt{d^2 + \frac{1}{4}v_\text{o}^2\Delta t^2}$ and the corresponding time in frame S

$$\Delta t = \frac{2}{c}\sqrt{d^2 + \frac{1}{4}v_\text{o}^2\,\Delta t^2}\ .$$

In order to relate the time Δt with $\Delta t'$, we first note that $d = d'$ (See Problem 8.1). Solving for Δt, we get

$$\Delta t^2 \left(1 - \beta^2\right) = \frac{4d^2}{c^2} = \Delta t'^2 \,, \tag{8.4}$$

where we have defined $\beta = v_o/c$. Thus, we see that time interval between the two events, the emission and the detection of light, as measured in frames S and S$'$, are not equal. Since both events occur at the same point in frame S$'$, we will denote S$'$ to be the *rest frame.* The time Δt in the moving frame S is related to the time $\Delta t'$ in the rest frame S$'$ by

$$\Delta t = \frac{4d^2}{c^2} = \frac{\Delta t'}{\sqrt{1-\beta^2}} \equiv \gamma\,\Delta t' \,. \tag{8.5}$$

Thus, the time interval between the two events in frame S will be greater than in the rest fame by a factor $1/\sqrt{1-\beta^2}$, which is commonly denoted by γ. This effect is known as *time dilation.*

8.3.2 Length contraction

We next consider sending a beam of light along a rod of length L' lying along the x'-axis in frame S$'$ as illustrated in Fig. 8.4. The light originates at one end of the bar, travels to the other end and is reflected back to its point of origin.

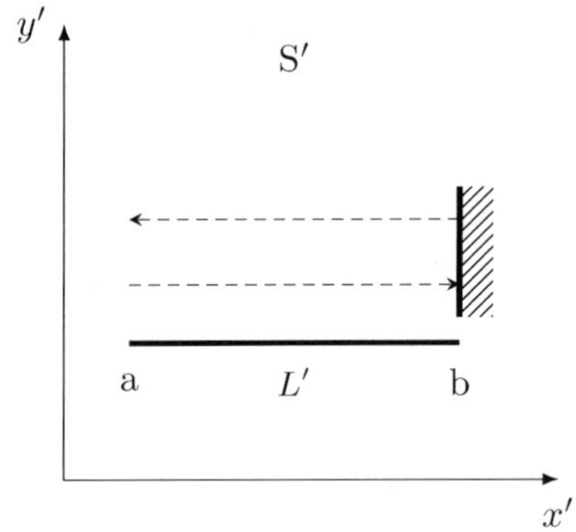

FIGURE 8.4
A bar of length L' lies along the x'-axis in frame S$'$. A light beam is sent from the end "a" reaches end "b" and is then reflected by a mirror.

We examine these events from the perspectives of observers in S and S$'$:

- Frame S$'$: As depicted in Fig. 8.4, since the bar is stationary in frame S$'$, the time for the round trip is $\Delta t' = 2L'/c$.

- Frame S: Fig. 8.5 describes this scenario. In this frame, the bar has length L and is moving with velocity $\vec{v}_o$ along the $+x$-axis. A light beam originates at point "a" and travels toward point "b." Since light moves at speed c in the x-direction, and the bar advances with speed v_o, the beam takes time $\Delta t_1 = L/(c - v_o)$ to reach point "b." Upon reflection at "b" it returns to point "a" in time $\Delta t_2 = L/(c + v_o)$. Thus, the total time between the two events is

$$\Delta t = \Delta t_1 + \Delta t_2 = \frac{L}{c - v_o} + \frac{L}{c + v_o} = \frac{2Lc}{c^2 - v_o^2} \; .$$

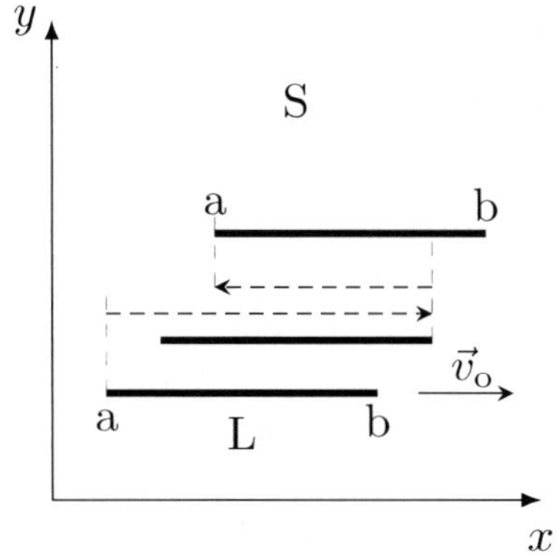

FIGURE 8.5
The figure portrays the bar of length L moving with a velocity $\vec{v}_o$ along the x-axis in frame S at three distinct times. The bottommost bar illustrates the position of the bar when a light beam emanates from point "a." The bar in the middle shows its position when the beam reaches point "b" and reflects back. Finally, the topmost bar represents the position when the beam returns to the point "a." Dashed horizontal lines represent the trajectory of the light beam towards the mirror and its journey back to point "a."

However, from Eq. (8.5), the time intervals Δt and $\Delta t'$ should be related by

$$\frac{2Lc}{c^2 - v_o^2} = \gamma \frac{2L'}{c} \; , \tag{8.6}$$

or

$$\gamma^2 L = \gamma L' \; ,$$

which gives

$$L = \frac{L'}{\gamma} \; . \tag{8.7}$$

This is known as the *length contraction.*

8.4 Lorentz transformation

We will now determine the linear equations relating space and time coordinates x', y', z', t' in frame S′ with coordinates x, y, z, t in frame S.

As we have stated before, the perpendicular dimensions remain unchanged; i.e., the corresponding transformations are $y' = y$ and $z' = z$. For the relation among coordinates in the direction of the relative velocity, we consider a point P′ situated on the x'-axis in frame S′. Its coordinates are $(x', 0, 0)$ in S′ and $(x, 0, 0)$ in S, respectively. The distance from P′ to the origin O′ as seen from frame S′ is x', and the same seen from frame S is $x - v_o t$. These two distances seen from two different frames are related by Eq. (8.7) as

$$x' = \gamma (x - v_o t) \ . \tag{8.8}$$

From symmetry, we should then have

$$x = \gamma (x' + v_o t') \ , \tag{8.9}$$

where we have reversed the sign of the velocity v_o. Using Eqs. (8.8) and (8.9), we arrive at

$$t' = \gamma \left(t - \frac{v_o}{c^2} x \right) \ .$$

Now, putting these all together, we get

$$\begin{aligned} x' &= \gamma (x - v_o t) \\ y' &= y \\ z' &= z \\ t' &= \gamma \left(t - \frac{v_o}{c^2} x \right) \ . \end{aligned} \tag{8.10}$$

This set of equations relating space-time coordinates in two inertial frames is known as *Lorentz Transformation.* The inverse set of transformations reads

$$\begin{aligned} x &= \gamma (x' + v_o t') \\ y &= y' \\ z &= z' \\ t &= \gamma \left(t' + \frac{v_o}{c^2} x' \right) \ . \end{aligned} \tag{8.11}$$

8.4.1 Velocity addition

The components (v_x, v_y, v_z) of the velocity of a particle in frames S are related to the components in S′ by

$$
\begin{aligned}
v_x = \frac{dx}{dt} &= \frac{dx' + v_o dt'}{dt' + \frac{v_o}{c^2}dx'} = \frac{v'_x + v_o}{1 + \frac{v_o v'_x}{c^2}} \\
v_y = \frac{dy}{dt} &= \frac{v'_y}{\gamma\left(1 + \frac{v_o v'_x}{c^2}\right)} \\
v_z = \frac{dz}{dt} &= \frac{v'_z}{\gamma\left(1 + \frac{v_o v'_x}{c^2}\right)} \ .
\end{aligned} \tag{8.12}
$$

In the non-relativistic limit, i.e., when $v_o \ll c$, we recover the Galilean velocity addition given in Eq. (8.2).

8.4.2 Four-vectors

We will now introduce several notations. We begin with the time and space coordinates for an event: (ct, x, y, z). Here, we have multiplied t by c to ensure all coordinates have dimensions of length. We denote these coordinates as (x^0, x^1, x^2, x^3). The transformations described by Eq. (8.10) can now be expressed in matrix form as

$$
\begin{pmatrix} x'^0 \\ x'^1 \\ x'^2 \\ x'^3 \end{pmatrix} = \begin{pmatrix} \gamma & -\gamma\beta & 0 & 0 \\ -\gamma\beta & \gamma & 0 & 0 \\ 0 & 0 & 1 & 0 \\ 0 & 0 & 0 & 1 \end{pmatrix} \begin{pmatrix} x^0 \\ x^1 \\ x^2 \\ x^3 \end{pmatrix} . \tag{8.13}
$$

This equation can be written as a matrix equation as

$$x' = \Lambda x \ , \tag{8.14}$$

or explicitly in terms of four-components as

$$x'^\mu = \sum_\nu \Lambda^\mu_\nu x^\nu \ , \tag{8.15}$$

where Λ^μ_ν are the μ, ν-elements of the matrix Λ. We employ the convention that the Greek letter indices run over $0, 1, 2, 3$ and Latin letter indices over $1, 2, 3$. The coefficients Λ^μ_ν are symmetric in the indices μ and ν. Off-diagonal elements indicate that we are working in a four-dimensional world where time and space are inter-related.

In general, we say that T^μ are the contravariant components of a *four-vector* T if they transform as

$$T'^\mu = \sum_\nu \Lambda^\mu_\nu T^\nu \ . \tag{8.16}$$

The inverse transformation is given by

$$T^{\mu} = \sum_{\nu} \left(\Lambda^{-1}\right)^{\mu}_{\nu} T'^{\nu} \,, \tag{8.17}$$

where the matrix Λ^{-1} is the inverse of Λ, and can be obtained from Λ by simply replacing β with $-\beta$.

8.4.2.1 Invariant distance

In our three-dimensional world, the Euclidean distance between two points, given by $dx^2 + dy^2 + dz^2$, remains invariant under arbitrary translations or rotations of coordinate axes. We will now determine a similar invariant for the four-dimensional world.

Since light moves with constant speed c in two inertial frames S and S$'$, if a light beam moves a distance dx in time dt is frame S and distance dx' in the corresponding time interval dt' in frame S$'$, then we must have

$$ds^2 \equiv c^2dt^2 - dx^2 = 0 \;\; ; \quad ds'^2 \equiv c^2dt'^2 - dx'^2 = 0 \,. \tag{8.18}$$

This raises a natural question: is ds^2 in general an invariant, for cases in which dx and dt represent the space and time intervals between two arbitrary events? For arbitrary dt, dx, dy, dz, we define *space-time interval* as

$$ds^2 = c^2dt^2 - dx^2 - dy^2 - dz^2 \,. \tag{8.19}$$

Using the Lorentz transformations Eq. (8.10), by direct calculation we find

$$ds^2 = ds'^2 \,. \tag{8.20}$$

Exercise 8.1 Prove Eq. (8.20).

8.4.2.2 Contravariant and covariant components

The quantities $x^{\mu} \equiv (x^0, x^1, x^2, x^3)$ are the contravariant components of a space-time event in S. We now define the *covariant components* of the same event as $x_{\mu} \equiv (x_0, x_1, x_2, x_3)$, where $x_0 = x^0, x_1 = -x^1, x_2 = -x^2, x_3 = -x^3$. Thus, the invariant distance ds^2 can be written as

$$ds^2 = \sum_{\mu} dx^{\mu} dx_{\mu} = \sum_{\mu,\nu} g_{\mu\nu} dx^{\mu} dx^{\nu} \,, \tag{8.21}$$

where the coefficients $g_{\mu\nu}$ are known as the *metric tensor*. They are given by $g_{\mu\nu} = 0$ if $\mu \neq \nu$ and $g_{00} = 1, g_{11} = g_{22} = g_{33} = -1$. The covariant components dx_{μ} are obtained from the contravariant ones via $dx_{\mu} = \sum_{\nu} g_{\mu\nu} dx^{\nu}$.

Exercise 8.2 If T^μ are the components of a four-vector, show that $\sum_\mu T_\mu T^\mu = \sum_{\mu,\nu} g_{\mu\nu} T^\nu T^\mu$ is Lorentz invariant.

The components of the *gradient operator* are defined as

$$\partial_\mu \equiv \frac{\partial}{\partial x'^\mu} \; ; \tag{8.22}$$

they transform like covariant components of a four-vector.

Exercise 8.3 Show that under Lorentz transformation the quantity $\sum_\mu dx^\mu \, \partial_\mu$ remains invariant; i.e.,

$$\sum_\mu dx^\mu \, \partial_\mu \to \sum_\mu dx'^\mu \, \partial'_\mu \; .$$

Transformation of covariant components

$$\begin{aligned} T'_\mu &= \sum_\nu g_{\mu\,\nu}\, T'^\nu = \sum_\nu g_{\mu\,\nu} \sum_\alpha \Lambda^\nu_\alpha \, T^\alpha \\ &= \sum_\nu g_{\mu\,\nu} \sum_\alpha \Lambda^\nu_\alpha \sum_\beta g^{\alpha\,\beta}\, T_\beta \; , \\ &= \sum_\beta \left(\sum_{\nu\,\alpha} g_{\mu\,\nu}\, \Lambda^\nu_\alpha \, g^{\alpha\,\beta} \right) T_\beta \; , \end{aligned} \tag{8.23}$$

where $g^{\alpha\,\beta} = g_{\alpha\,\beta}$. We have arrived at the following identity

$$\sum_{\mu\,\nu\,\alpha} \Lambda^\mu_\gamma \, g_{\mu\,\nu}\, \Lambda^\nu_\alpha \, g^{\alpha\,\beta} = \delta^\beta_\gamma \; ,$$

or, in matrix form

$$g\,\Lambda\, g = \Lambda^{-1} \; . \tag{8.24}$$

8.4.2.3 Four-velocity and four-momentum

We will now associate a four-vector with the velocity of a particle. Consider a particle moving with a velocity $\vec{v}$ in frame S. The velocity of the particle in a co-moving frame S′ will be zero. So, if the displacement of the particle in frame S during the time-interval dt is (dx, dy, dz), the identity $ds^2 = ds'^2$ leads to

$$c^2 dt^2 - dx^2 - dy^2 - dz^2 = c^2 dt'^2 \; .$$

Thus, the time dt' in the rest frame of the particle is a Lorentz invariant, known as the *proper time*, and denoted by $d\tau$.

Any four-vector divided by $d\tau$ would also be a four-vector. Since $(c\,dt, dx, dy, dz)$ are components of a four-vector, we now define the *four-velocity* of the particle to be

$$u^\mu = (c\,dt, dx, dy, dz)/d\tau \equiv (\gamma c, \gamma\vec{v}) \; . \tag{8.25}$$

Exercise 8.4 Show that

$$\frac{dt}{d\tau} = \gamma \; , \qquad \frac{\vec{i}\,dx + \vec{j}\,dy + \vec{k}\,dz}{d\tau} = \gamma\vec{v} \; .$$

Since u^μ is a four-vector, the quantity $\sum_\mu u^\mu u_\mu$ is an invariant:

$$\sum_\mu u^\mu u_\mu = c^2 \; . \tag{8.26}$$

Exercise 8.5 Prove Eq. (8.26).

We define the *four-momentum* p of the particle by

$$p^\mu = mu^\mu \; , \tag{8.27}$$

where m is the mass of the particle. The four components of p are

$$p \equiv (p^0, \vec{p}) = (m\gamma c, m\gamma\vec{v}) \; . \tag{8.28}$$

In the non-relativistic limit ($\gamma \to 1$), the three components p^1, p^2 and p^3, go into the usual three components of the momentum vector. To understand the nature of the zeroth component p^0, we expand it for small v, i.e., $v \ll c$. We get

$$p^0 = m\gamma c = \frac{m\,c}{\sqrt{1 - \frac{v^2}{c^2}}} \simeq m\,c + \frac{1}{2c} m\,v^2 \; .$$

Thus, we have

$$c\,p^0 = m\gamma c^2 \simeq m\,c^2 + \frac{1}{2} m\,v^2 \; .$$

Hence at low speeds, as the speed v of the particle changes, the change of the cp^0-term is equal to the change in kinetic energy. This suggest to identify the energy of the particle as

$$E = c\,p^0 \; . \tag{8.29}$$

Thus, energy and momentum form the four components of the momentum four-vector p:

$$p = (E/c, \vec{p}) \; . \tag{8.30}$$

From Eq. (8.26), we have

$$\sum_{\mu} p^{\mu} p_{\mu} = E^2/c^2 - \vec{p}^{\,2} = m^2 c^2 ,$$

and hence

$$E^2 = \vec{p}^{\,2} c^2 + m^2 c^4 . \tag{8.31}$$

8.4.3 Light cone

Consider a flash of light moving with speed c along the x-axis in an inertial frame. In the x^0-x^1 plane, this motion is described by a straight line that makes an angle of 45° with the x-axis in the first quadrant as shown in Fig. 8.6. The motion of a massive particle would be described a by a curve, $x(t) = v(t)t$. The tangents at any point of this curve would make an angle higher than 45°, because $v(t) < c$. The origin corresponds to the present. All points on the curve in the upper cone defined by $x^0 > x^1$ and $x^0 > -x^1$, describe the future of the particle. Similarly, the past of the particle lies in the lower cone

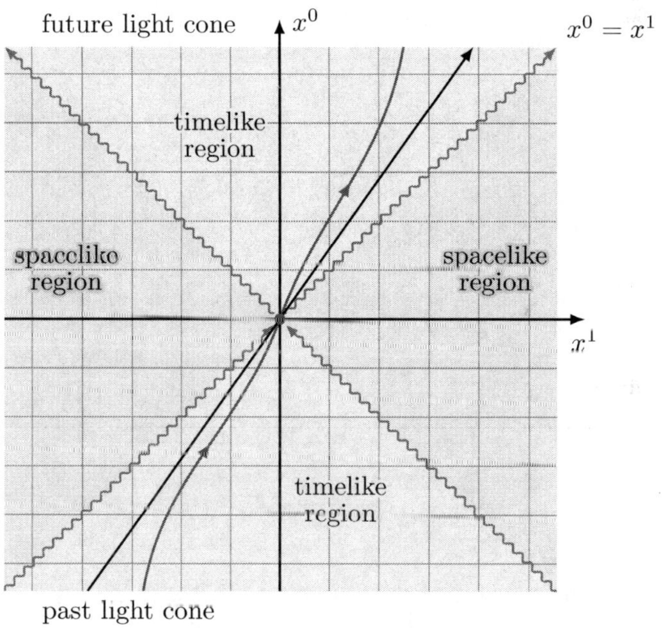

FIGURE 8.6
The figure shows the world lines of two particles, one moving with a constant velocity v and the other with $v(t)$. The dashed lines depict light rays and are necessarily at 45° to both x^0 and x^1 axes.

described by $x^0 < x^1$ and $x^0 < -x^1$. The locus traversed by the particle in the past and to be occupied in future is known as the *world line* of the particle. The cone delimited by $x^0 = \pm ct$ is known as the *light cone*. It is easily generalized to the case of three space and one time coordinates.

Exercise 8.6 Reason why the locus of the light pulse would be described by a straight line inclined at 45° to the x-axis.

8.4.4 Electrodynamics and Special Relativity

8.4.4.1 Current four-vector

Conservation of electric charge is a foundational principle in physics validated by all experiments over the past century.

Since the total charge is conserved, the charge densities will appear different in two inertial frames. In the rest frame S′ the charge enclosed in the volume dV' is

$$dQ = \rho'(x', y', z')\, dV' \ .$$

In frame S, the same charge is

$$dQ = \rho(x, y, z)\, dV \ .$$

Since $dV = dV'/\gamma$, the charge in the two frames are related by

$$\rho(x, y, z) = \gamma\, \rho'(x', y', z') \ .$$

To see how current densities transform under Lorentz transformations, we note that conservation of charge implies that continuity equations must hold in both S and S′. I.e.,

$$\frac{\partial \rho'}{\partial t'} + \vec{\nabla}' \cdot \vec{J}' = \frac{\partial \rho}{\partial t} + \vec{\nabla} \cdot \vec{J} = 0 \ . \tag{8.32}$$

In other words, $\frac{\partial \rho}{\partial t} + \vec{\nabla} \cdot \vec{J}$ must be a Lorentz invariant. If we define a four-component quantity $J^\mu \equiv (c\rho, \vec{J})$, we can write Eq. (8.32) as

$$\sum_\mu \frac{\partial J^\mu}{\partial x^\mu} \equiv \sum_\mu \partial_\mu J^\mu = 0 \ , \tag{8.33}$$

where ∂_μ are the components of the gradient operator of Eq. (8.22) and transform like covariant components of a four-vector.

Since Eq. (8.33) is valid for arbitrary charge and current densities in any inertial frame, the quantity $J^\mu \equiv (c\rho, \vec{J})$ must transform like the contravariant components of a four-vector, known as the *current four-vector*.

8.4.4.2 Potential four-vector

We have seen that the gradient ∂_μ transforms as covariant components of a four-vector. We now define $\partial^\mu = \sum_\mu g^{\mu\nu}\partial_\nu$. Thus: $\partial^0 = \partial_0$, $\partial^1 = -\partial_1, \partial^2 = -\partial_2$, and $\partial^3 = -\partial_3$. Computing the sum $\sum_\mu \partial^\mu\partial_\mu$, we obtain

$$\sum_\mu \partial^\mu \partial_\mu = \frac{1}{c^2}\frac{\partial^2}{\partial t^2} - \frac{\partial^2}{\partial x^2} - \frac{\partial^2}{\partial y^2} - \frac{\partial^2}{\partial z^2} \, . \tag{8.34}$$

As indicated in Prob. 8.3, the operator $\sum_\mu \partial^\mu\partial_\mu$ is an invariant, and this implies that operators ∂^μ must transform as contravariant components of a four-vector.

In terms of the operator $\sum_\mu \partial^\mu\partial_\mu$, Eqs. (6.17) and (6.18) can be written as

$$\sum_\mu \partial^\mu \partial_\mu \, A^\nu = \; - \; \mu_o \, J^\nu \; , \tag{8.35}$$

where we have defined the quantities A^μ by

$$A^\mu \equiv \left(V/c, \vec{A}\right) \, . \tag{8.36}$$

Eq. (8.35) is valid independent of the space-time point or the reference frame in which it is computed, which implies that the both sides of the equation must transform in the same manner under Lorentz transformations. Since the right-hand side represents a four-vector, the left-hand side must transform as a four-vector as well. With the Lorentz invariance of $\sum_\mu \partial^\mu\partial_\mu$, we see that A^μ must transform as contravariant components of a four-vector, which is known as *four-potential.*

We find further corroboration of the four-vector nature of A^μ from the Lorentz gauge condition, Eq (6.15):

$$\vec{\nabla} \cdot \vec{A} + \frac{1}{c^2}\frac{\partial V}{\partial t} = 0,$$

which can now be written as

$$\sum_\mu \partial_\mu A^\mu = 0. \tag{8.37}$$

Thus, unlike the Coulomb gauge, the Lorentz gauge condition is manifestly Lorentz invariant.

8.4.4.3 Einstein's summation convention

Having seen several examples of four-vectors and Lorentz invariants, we now introduce the Einstein's summation convention that will make our expressions appear much less cluttered. When an index is repeated within a term, we will automatically assume summation over that index and do not write the summation sign.

With this convention, notations simplify considerably as shown below:

$$\sum_\mu dx^\mu dx_\mu \equiv dx^\mu dx_\mu, \ \sum_\mu \partial_\mu A^\mu \equiv \partial_\mu A^\mu, \ \sum_\mu \partial_\mu J^\mu \equiv \partial_\mu J^\mu \ .$$

In each of these summations, μ is the repeated index, appearing once as a covariant index and once as a contravariant index in every term.

Using Einstein's summation convention, the Eqs. (8.33), (8.35), and (8.37) can now be written respectively as

$$\partial_\mu J^\mu = 0 \ , \quad \partial^\mu \partial_\mu A^\nu = \ - \ \mu_o J^\nu \ , \quad \partial_\mu A^\mu = 0 \ .$$

The index ν appears in both sides of the equation and is known as a free index.

8.4.4.4 Field tensor

So far we have encountered physical quantities that are Lorentz invariant such as ds^2 that have no indices, or four-vectors such as J^μ, A_μ with one index. There are quantities which transform under Lorentz transformations as components of a second rank tensor, i.e., like a product of two four-vectors:

$$F'^{\mu\nu} = \Lambda^\mu_\alpha \, F^{\alpha\beta} \, \Lambda^\nu_\beta \ . \tag{8.38}$$

We will now show that the $\vec{E}$ and $\vec{B}$ fields are components of a second rank tensor, the *electromagnetic field tensor*.

The electric field is given by

$$\vec{E} = -\vec{\nabla} V - \frac{\partial \vec{A}}{\partial t} \ ,$$

which can be written as

$$\vec{E} = -\vec{\nabla}(c\,A^0) - c\,\frac{\partial \vec{A}}{\partial x^0} \ . \tag{8.39}$$

I.e., the i-th component of the electric field E^i is

$$\begin{aligned} E^i &= -\partial_i (c\,A^0) - (c\,\partial_0) A^i \\ &= -(-\partial^i)(c\,A^0) - (c\,\partial^0) A^i \\ &= -c\left(\partial^0 A^i - \partial^i A^0\right) \ . \end{aligned} \tag{8.40}$$

In a similar manner, using the Einstein summation convention, the components of the magnetic field $\vec{B} = \vec{\nabla} \times \vec{A}$ can be written as

$$B^i \ = \ -\epsilon^i_{jk} \, \partial^j A^k \ , \tag{8.41}$$

where $\epsilon^i_{jk} \equiv \epsilon_{ijk}$ are the components of the Levi-Civita tensor. We define a second rank antisymmetric tensor $F^{\mu\nu}$

$$F^{\mu\nu} = \partial^\mu A^\nu - \partial^\nu A^\mu \ , \tag{8.42}$$

whose components are related to $\vec{E}$ and $\vec{B}$ by

$$F^{0i} = -E_i/c \; ; \qquad F^{ij} = -\epsilon^{ij}_{k} B_k \; , \tag{8.43}$$

where $\epsilon^{ij}_{k} \equiv \epsilon_{ijk}$.

Exercise 8.7 Prove Eq. (8.43).

In matrix form, the components of the electromagnetic field tensor $F^{\mu\nu}$ can be written as

$$\mathcal{F} \equiv \begin{bmatrix} 0 & -\frac{1}{c}E_1 & -\frac{1}{c}E_2 & -\frac{1}{c}E_3 \\ \frac{1}{c}E_1 & 0 & -B_3 & B_2 \\ \frac{1}{c}E_2 & B_3 & 0 & -B_1 \\ \frac{1}{c}E_3 & -B_2 & B_1 & 0 \end{bmatrix} . \tag{8.44}$$

There is another second rank tensor $G^{\mu\nu}$, which can be obtained from $F^{\mu\nu}$ by the following replacements:

$$\vec{E} \to c\vec{B} \; , \qquad \vec{B} \to -\frac{\vec{E}}{c} \; .$$

In matrix form, the tensor $G^{\mu\nu}$ is given by

$$\mathcal{G} \equiv \begin{bmatrix} 0 & -B_1 & -B_2 & -B_3 \\ B_1 & 0 & \frac{1}{c}E_3 & -\frac{1}{c}E_2 \\ B_2 & -\frac{1}{c}E_3 & 0 & \frac{1}{c}E_1 \\ B_3 & \frac{1}{c}E_2 & -\frac{1}{c}E_1 & 0 \end{bmatrix} . \tag{8.45}$$

The tensor $G^{\mu\nu}$ is known as the *dual* to tensor $F^{\mu\nu}$. The product of the two tensors is Lorentz invariant:

$$G^{\mu\nu} F_{\mu\nu} - \frac{4}{c} \left(\vec{B} \cdot \vec{E} \right) \; , \tag{8.46}$$

where we have used Einstein summation convention.

We will now write the four Maxwell's equations using the tensors $F^{\mu\nu}$ and $G^{\mu\nu}$. The homogeneous Maxwell's equations

$$\vec{\nabla} \times \vec{E} + \frac{\partial \vec{B}}{\partial t} = 0 \; , \qquad \vec{\nabla} \cdot \vec{B} = 0 \; ,$$

can be written as

$$\partial_\mu G^{\mu\nu} = 0 \; . \tag{8.47}$$

Similarly, the inhomogeneous equations with source densities

$$\vec{\nabla} \cdot \vec{E} = \rho/\epsilon_o \ , \qquad \vec{\nabla} \times \vec{B} = \mu_o \left(\vec{J} + \epsilon_o \frac{\partial \vec{E}}{\partial t} \right) \ , \tag{8.48}$$

become

$$\partial_\mu F^{\mu\nu} = \mu_o J^\nu \ . \tag{8.49}$$

Exercise 8.8 Show that Eqs. (8.47) and (8.49) correctly reproduce the four Maxwell's equations.

Example 8.1 EM field due to a cylindrical distribution of charge
An infinitely long cylindrical charge distribution is symmetrically placed along the x'-axis in frame S$'$ and is described by the time-independent charge density

$$\rho(r', \theta', x') = \sigma_o \, \delta\,(r' - R) \ ; \quad \sigma_o > 0 \ ,$$

where $r' = \sqrt{y'^2 + z'^2}$ and $\theta' = \tan^{-1}(z'/y')$.

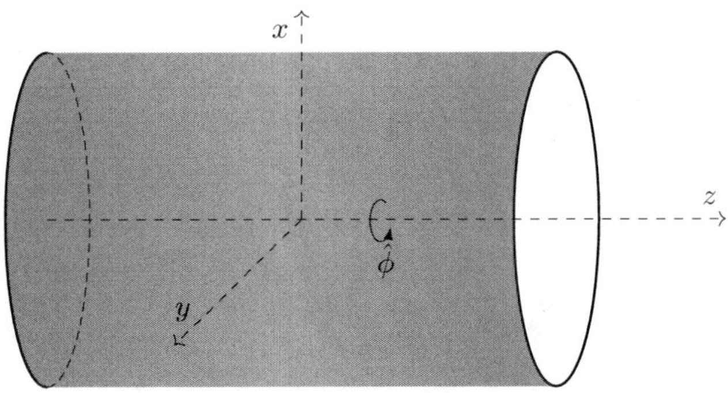

FIGURE 8.7
A long cylindrical shell symmetrically placed around the current which flows along the z-axis.

- Determine the electric field $\vec{E}'$ and the field tensor $F'^{\mu\nu}$ at the space-time point $(ct', 0, y', 0)$, where $y' > R$.

  Solution: Let us consider a cylindrical Gaussian surface of radius $r' > R$ and length ℓ'. Because the electric field is radially outward from the x-axis, the flux out of this surface is $2\pi r' \ell' \, E'_r$, which should be equal to q_{in}/ε_o, where q_{in} is the charge enclosed by the surface

$$q_{in} = \int_0^{r'} dr' \int_0^{\ell'} dx' \int_0^{2\pi} r' d\theta' \, \rho(r', \theta', x') = 2\pi R \, \ell' \, \sigma_o \ .$$

Thus, using Gauss' law, we get

$$\vec{E}' = \frac{R}{r'}\frac{\sigma_o}{\varepsilon_o}\,\hat{e}_{r'} \;,$$

which at the space-time point $(ct', 0, y', 0)$ becomes

$$\vec{E}' = \left(0\,,\;\frac{R}{r'}\frac{\sigma_o}{\varepsilon_o}\,,\;0\right)\,.$$

Because the magnetic field is zero in S′, the tensor $F'^{\mu\nu}$ can be written in matrix form

$$\mathcal{F}' = \begin{bmatrix} 0 & 0 & -\frac{1}{c}\frac{R}{r'}\frac{\sigma_o}{\varepsilon_o} & 0 \\ 0 & 0 & 0 & 0 \\ \frac{1}{c}\frac{R}{r'}\frac{\sigma_o}{\varepsilon_o} & 0 & 0 & 0 \\ 0 & 0 & 0 & 0 \end{bmatrix}\,. \tag{8.50}$$

- Find the field tensor $F^{\mu\nu}$ and fields $\vec{E}$ and $\vec{B}$ at the corresponding space-time point $(ct, 0, y, 0)$ in frame S, and generalize it for an arbitrary point.

  Solution: Inverting Eq. (8.38), the field tensor $\mathcal{F}$ is given by

$$\mathcal{F} = \Lambda^{-1}\,\mathcal{F}'\,\Lambda^{-1}\;, \tag{8.51}$$

where

$$\Lambda^{-1} = \begin{pmatrix} \gamma & \gamma\beta & 0 & 0 \\ \gamma\beta & \gamma & 0 & 0 \\ 0 & 0 & 1 & 0 \\ 0 & 0 & 0 & 1 \end{pmatrix}\,.$$

Then, the matrix $\Lambda^{-1}\,\mathcal{F}'\,\Lambda^{-1}$ becomes

$$\mathcal{F} \;=\; \begin{bmatrix} 0 & 0 & -\frac{\gamma}{c}\frac{R}{r'}\frac{\sigma_o}{\varepsilon_o} & 0 \\ 0 & 0 & -\frac{\gamma\beta}{c}\frac{R}{r'}\frac{\sigma_o}{\varepsilon_o} & 0 \\ \frac{\gamma}{c}\frac{R}{r'}\frac{\sigma_o}{\varepsilon_o} & \frac{\gamma\beta}{c}\frac{R}{r'}\frac{\sigma_o}{\varepsilon_o} & 0 & 0 \\ 0 & 0 & 0 & 0 \end{bmatrix}$$

Thus, from Eq. (8.44), we identify the field components in S, at the space-time point $(ct, 0, y, 0)$:

$$\vec{E} = \gamma\frac{R}{r}\frac{\sigma_o}{\varepsilon_o}\,\hat{e}_y\;, \quad \vec{B} = \frac{\gamma\beta}{c}\frac{R}{r}\frac{\sigma_o}{\varepsilon_o}\,\hat{e}_z\;.$$

Generalizing this result for an arbitrary point in frame S, we get

$$\vec{E} = \gamma \frac{R}{r} \frac{\sigma_o}{\varepsilon_o} \hat{r} \,, \quad \vec{B} = \frac{\gamma\beta}{c} \frac{R}{r} \frac{\sigma_o}{\varepsilon_o} \hat{\varphi} \,,$$

where $\hat{r}$ is a unit vector that points away from the x-axis and $\hat{\varphi}$ is the unit vector related angular coordinate φ that wraps around the x-axis in a right-handed manner.

- Determine $\vec{E}$ and $\vec{B}$ fields using Maxwell's equations and transformed source distribution in frame S.

  Solution: The four-current density in frame S′ is

$$J' = \left(c\sigma_o \,\delta\left(r' - R\right), 0\,, 0\,, 0\right) \,.$$

  Hence, the corresponding four-current density J in frame S is $J = \Lambda^{-1} J'$, or

$$J = \begin{pmatrix} \gamma & \gamma\beta & 0 & 0 \\ \gamma\beta & \gamma & 0 & 0 \\ 0 & 0 & 1 & 0 \\ 0 & 0 & 0 & 1 \end{pmatrix} \begin{pmatrix} c\sigma_o \,\delta\left(r' - R\right) \\ 0 \\ 0 \\ 0 \end{pmatrix} = \begin{pmatrix} c\gamma\sigma_o \,\delta\left(r - R\right) \\ c\gamma\beta\sigma_o \,\delta\left(r - R\right) \\ 0 \\ 0 \end{pmatrix}$$

  From Gauss' and Ampere's laws the fields in frame S are then given by $\vec{E} = \gamma \frac{R}{r} \frac{\sigma_o}{\varepsilon_o} \hat{r}$ and $\vec{B} = \frac{\gamma\beta}{c} \frac{R}{r} \frac{\sigma_o}{\varepsilon_o} \hat{\varphi}$.

8.5 Problems

8.1 Assuming the space to be isotropic, reason why the lateral dimensions, the dimensions perpendicular to the relative velocity $\vec{v}_o$, must remain unaffected as a part of coordinate transformation. That is, the transformation equations between frames S and S′ must include

$$\Delta y' = \Delta y; \qquad \Delta z' = \Delta z \,.$$

Hint: Consider from the perspective of an observer in an inertial frame S_o that sees two inertial frames $S_\pm$ move with velocities $\pm\vec{v}_o$ towards it.

8.2 For two four-vectors A^μ and B^μ that transform as $T'^\mu = \sum_\nu \Lambda^\mu_\nu T^\nu$, show that $\sum_{\mu,\nu} g_{\mu\nu} A^\mu B^\nu$ is Lorentz invariant.

8.3 In Eq. (4.13), we saw that, in absence of sources, both electric and magnetic fields satisfy the wave equation, which for one dimension reduces to

$$\left(\frac{\partial^2}{\partial x^2} - \frac{1}{c^2}\frac{\partial^2}{\partial t^2}\right)\begin{pmatrix}\vec{E}\\ \vec{B}\end{pmatrix} = 0 , \tag{8.52}$$

where $c = \frac{1}{\sqrt{\mu_o \epsilon_o}} = 3 \times 10^8\,\mathrm{m\,s^{-2}}$. Under Galilean transformations

$$t' = t \; ; \quad x' = x - v_o t , \tag{8.53}$$

the wave equation does not retain its form, but it does so under Lorentz transformation given in Eq. (8.10).

8.4 Prove Eq. (8.46). This shows that $\vec{E}\cdot\vec{B}$ is a Lorentz Invariant.

8.5 Show that

$$F^{\mu\nu}F_{\mu\nu} = 2\left(E^2/c^2 - B^2\right) ,$$

and thus establish that $c^2B^2 - E^2$ is a Lorentz Invariant.

8.6 In Chap. 6, we saw that general gauge transformations

$$\vec{A} \to \vec{A}' = \vec{A} + \nabla\chi \; , \; V \to V' = V - \frac{\partial\chi}{\partial t},$$

leave electromagnetic fields invariant.

(a) Show that these transformations can be succinctly written as

$$A_\mu \to A'_\mu = A_\mu + \partial_\mu\chi \; ;$$

(b) the field tensors $F_{\mu\nu}$ are invariant under $A_\mu \to A'_\mu = A_\mu + \partial_\mu\chi$.

8.7 Two large parallel plates of area A are placed parallel to the x'-y' plane in the reference frame S′. They carry charges $\pm\sigma_o$ per unit area.

- Determine the electric field $\vec{E}'$ and the field tensor $F'^{\mu\nu}$ at a space-time point whose spatial components lie between the plates
- Determine the field tensor $F^{\mu\nu}$ and fields $\vec{E}$ and $\vec{B}$ at the corresponding space-time point $(x^0, \vec{x})$ in frame S.
- Justify your answers from Maxwell's equations and source distribution in S.

Part III

Quantum Mechanics

9

Quantum Mechanics: A Brief Review

In this chapter we will briefly review the fundamental ideas of quantum mechanics. We assume that the reader took an introductory course in quantum mechanics, so this section is intended primarily to establish notation.

9.1 Introduction

In classical physics, we are used with deterministic and "precise" laws stemming from the fact that the act of measurement could be ignored. Indeed, for all practical purposes, the interaction with the macroscopic system to be measured can be rendered negligibly small. However, at very small scales, the situation changes radically. For micro-objects, the assumption that the measurement is ignorable, is no longer true. In order to perform a measurement we must interact with the system we are measuring. If we could refine our measurements indefinitely, and thus, reduce our interaction arbitrarily, then the notion of big or small would make no sense [11], for we could leave relatively unperturbed any system, regardless of their scale. However, we discovered experimentally that the nature has an objective scale, encoded by Planck's constant, that limits our ability to perform perturbation-free measurements. In this realm all we can do is to perform a large number of measurements on identically prepared systems, which would allow us to determine the probability of a given outcome. We found that at very small scales, micro-objects behave in a non-intuitive way, described by new laws, those of quantum mechanics, which has been successful in calculating very precisely these probabilities. However, quantum mechanics cannot predict a particular outcome of a single measurement. Thus, the classical deterministic way of looking at nature no longer applies.

DOI: 10.1201/9781003174967-9

9.2 Postulates of quantum mechanics

In quantum mechanics the state of a system is described by a vector in a suitably chosen complex Hilbert space $\mathbb{H}$. We denote the *inner product* of two vectors in $\mathbb{H}$ by $\langle A|B\rangle$. The inner is in general a complex number; i.e., $\langle A|B\rangle \in \mathbb{C}$, and it satisfies $\langle A|B\rangle = \overline{\langle B|A\rangle}$, where the bar denotes complex conjugation. It also satisfies $\langle A|A\rangle \geq 0$, with the equal sign only when $|A\rangle = 0$. This will be used to normalize any vector $|A\rangle$ by multiplying it with $\langle A|A\rangle^{-1/2}$. The inner product is linear in the second argument.

We will use Dirac's ket notation [11] and denote the states of the system by the *ket* symbol $|\ \rangle$. If $|A\rangle$ and $|B\rangle$ are two ket vectors, then their linear superposition

$$c_1 |A\rangle + c_2 |B\rangle = |R\rangle \ , \tag{9.1}$$

is also a ket vector. Here c_1 and c_2 are two complex constants. If $|\alpha\rangle$ is a ket vector depending on a variable α that takes all values in a certain range, then we can integrate with respect to α and get another ket vector

$$\int |\alpha\rangle \, d\alpha = |Q\rangle \ . \tag{9.2}$$

Postulate 1 (States)
The state of a system is represented by a normalized ket vector $|\Psi\rangle$ belonging to a Hilbert space $\mathbb{H}$ called the state space.

To any ket vector we associate a unique conjugate vector which we call a *bra vector* and denote by $\langle\ |$. Bra vectors live in the dual space $\mathbb{H}^*$. When acting on the ket vectors they yield complex numbers. We define the action of the bra $\langle A|$ on an arbitrary ket $|B\rangle$ such that

$$\langle A| \left(|B\rangle \right) = \langle A|B\rangle \ . \tag{9.3}$$

In other words, we define $\langle A|$ such that its action on $|B\rangle$ gives the same number as the inner product of the ket vectors $|A\rangle$ and $|B\rangle$. This definition gives the sought after one-to-one correspondence $\langle A| \leftrightarrow |A\rangle$. Note that while ket vectors live in the Hilbert space $\mathbb{H}$, their bra conjugates live in the dual space $\mathbb{H}^*$, as illustrated in Fig. 9.1.

The bra-ket notation allows for elegant expressions for linear operators. A linear operator is a linear map from one vector space to another. In our case it maps one vector $|A\rangle$ from the Hilbert space $\mathbb{H}$ to another ket vector in the same space. Some linear operators can be elegantly written as $|B\rangle\langle C|$. Indeed,

$$\left(|B\rangle \langle C| \right) |A\rangle \equiv |B\rangle \langle C|A\rangle = \langle C|A\rangle \, |B\rangle = |D\rangle \ , \tag{9.4}$$

where we used the fact that $\langle C|A\rangle$ is just a complex number, and denoted $|D\rangle = \langle C|A\rangle \, |B\rangle$.

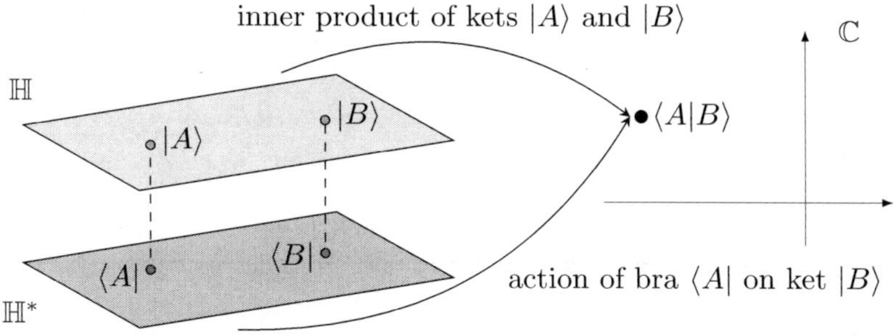

FIGURE 9.1
The Hilbert space $\mathbb{H}$ and its dual $\mathbb{H}^*$. The inner product of kets $|A\rangle$ and $|B\rangle$ and the action of bra $\langle A|$ on ket $|B\rangle$ both give the same complex number denoted by $\langle A|B\rangle$.

A projection operator P_A, maps a vector $|B\rangle$ into another vector along the direction of $|A\rangle$. Using the bra-ket formalism, the projection operator can be expressed as $P_A = |A\rangle\langle A|$. We immediately see that when P_A acts on a vector $|B\rangle$ gives the component of $|B\rangle$ which is along $|A\rangle$

$$P_A|B\rangle = |A\rangle\langle A|B\rangle \tag{9.5}$$

If $|A\rangle$ and $|B\rangle$ are orthogonal, then $\langle A|B\rangle = 0$ and the projection operator yields, as expected, zero.

> **Exercise 9.1** Show that not all linear operators can be written in the universal ket-bra form. *Hint:* Think for example, about rotation operators.

If a discrete set of ket vectors $\{|n\rangle\}$ is complete, then any other ket vector $|\Psi\rangle$ in the Hilbert space can be expressed as a linear combination

$$|\Psi\rangle = \sum_n c_n |n\rangle \ . \tag{9.6}$$

Similarly, for the continuous case, the set $\{|q\rangle\}$ is complete if

$$|\Psi\rangle = \int c(q)\,|q\rangle\, dq \ . \tag{9.7}$$

A set of vectors is orthonormal if

$$\langle m|n\rangle = \delta_{m,n} \qquad \text{for the discrete case} \tag{9.8}$$

$$\langle q|q'\rangle = \delta(q-q') \qquad \text{for the continuum case} \tag{9.9}$$

where $\delta_{m.n}$ is the Kronecker symbol and $\delta(q)$ is Dirac delta-function. If a set of kets is orthonormal and complete then their corresponding projection operators satisfy

$$\sum_n |n\rangle\langle n| = \mathbf{1} \quad \text{for the discrete case} \tag{9.10}$$

$$\int |q\rangle\langle q|\, dq = \mathbf{1} \quad \text{for the continuum case} \tag{9.11}$$

where **1** is the identity operator. Equations (9.10) and (9.11) are an equivalent way of expressing the completeness of a set of orthonormal kets.

Exercise 9.2 Prove Eqs. (9.10) and (9.11).

Physical quantities are described by Hermitian operators. We associate a Hermitian operator $\hat{x}$ to position x and $\hat{p}$ to momentum p. These operators obey the commutation relation

$$[\hat{x}, \hat{p}] \equiv \hat{x}\hat{p} - \hat{p}\hat{x} = i\hbar\,\mathbf{1}\ . \tag{9.12}$$

Henceforth, in order to simplify notation, we will omit writing explicitly the identity operator **1**, and later on, when there is no danger of confusion, we will drop the hat notation for the operators as well.

In general, to a physical quantity $Q(x, p)$ we associate a Hermitian operator $\widehat{Q}(\hat{x}, \hat{p})$ where we replaced the position and momentum by the corresponding operators. This quantization scheme works well, with the caveat that special attention should be paid for cases were we have products of x and p, because their quantum counterparts do not commute. For example, in quantum mechanics xp is replaced by $(\hat{x}\hat{p} + \hat{p}\hat{x})/2$ where the symmetrization is necessary to avoid ambiguity. We will not go into the more subtle points of quantization, because they are beyond the scope of this book.

Postulate 2 (Observables)
Observable physical quantities like position, momentum, energy, etc., are represented by Hermitian operators in a Hilbert space.

For example, the Hamiltonian of a particle of mass m subject to a force generated by the potential $V(x)$ is given by

$$\widehat{H} = \frac{\hat{p}^2}{2m} + V(\hat{x})\ . \tag{9.13}$$

Hermitian operators have real eigenvalues. That is, if $|q\rangle$ is an eigenvector of a Hermitian operator $\hat{Q}$, then

$$\hat{Q}\,|q\rangle = q\,|q\rangle\ , \quad q \in \mathbb{R}\ . \tag{9.14}$$

This property is important in connecting the theory with the experiment because the quantities we measure are real numbers. Additionally, the eigenvectors of Hermitian operators form a complete set, and the eigenvectors corresponding to two different eigenvalues are orthogonal.

We postulate that when we measure a given physical quantity, we obtain only the values given by the set of the eigenvalues (the spectrum) of the corresponding Hermitian operator. An observable can have a discrete spectrum, a continuous spectrum, or a combination of the two. For example, for a particle in the infinite square well, the Hamiltonian has a discrete spectrum; i.e., the particle can have only discrete energies. By contrast, a free particle has a continuous energy spectrum. There are systems that can have both continuous and discrete spectra such as the electron in the hydrogen atom. It has discrete spectrum for the bound states, and continuous spectrum when the atom is ionized.

Postulate 3 (Measurement)
If a system is in state $|\Psi\rangle$ and we measure an observable $\hat{Q}$, we obtain a real number q which is one of the eigenvalues of $\hat{Q}$. As a result of the measurement, the system transitions from $|\Psi\rangle$ to $|q\rangle$, where $|q\rangle$ is the corresponding eigenvector of $\hat{Q}$. The probability of measuring q is

$$P(q) = |\langle q|\Psi\rangle|^2 \tag{9.15}$$

for the case in which the spectrum of $\hat{Q}$ is discrete, and

$$dP(q) = |\langle q|\Psi\rangle|^2 dq \tag{9.16}$$

for the continuous spectrum.

The last postulate concerns the evolution of a physical state $|\Psi(t)\rangle$. This is also known as Schrödinger equation.

Postulate 4 (Evolution)
Consider a system described by the Hamiltonian $\hat{H}$, that is in state $|\Psi(t)\rangle$. Left unperturbed, the system's time evolution is given by the Schrödinger equation

$$i\hbar\frac{\partial}{\partial t}|\Psi(t)\rangle = \hat{H}|\Psi(t)\rangle \ . \tag{9.17}$$

In solving concrete physical problems, it is convenient to work in a given basis of the Hilbert space $\mathbb{H}$. For example, in the one-dimensional case, we denote by $\Psi(x,t)$ the projection of $|\Psi(t)\rangle$ in basis $|x\rangle$; i.e., $\Psi(x,t) = \langle x|\Psi(t)\rangle$. $\Psi(x,t)$ is called the wave function of the system. Then, the Schrodinger equation (9.17) becomes

$$i\hbar\frac{\partial}{\partial t}\Psi(x,t) = \hat{H}\Psi(x,t) \ , \tag{9.18}$$

where $\hat{H} = -\frac{\hbar^2}{2m}\frac{\partial^2}{\partial x^2} + V(x)$.

In many cases of practical interest, the Hamiltonian is independent of time. Then, the wave function separates

$$\Psi(x,t) = e^{-\frac{i}{\hbar}Et}\psi_E(x) \tag{9.19}$$

where $\psi_E(x)$ is the eigenfunction of the Hamiltonian $\hat{H}$ corresponding to the eigenenergy E

$$\hat{H}\psi_E(x) = E\psi_E(x) \; . \tag{9.20}$$

Equation (9.20) is known as the time independent Schrödinger equation. The generalization to the three-dimensional case is straightforward.

9.3 Pictures of quantum mechanics

In the formalism we have described up to this point, a quantum state is a vector in the Hilbert space of quantum mechanical states, which evolves in time according to Schrödinger equation. This is known as the *Schrödinger picture or representation*, in which operators are fixed, while the physical states evolve in time. There is another, equivalent, way of describing the evolution of the systems, known as the *Heisenberg picture*, where operators are time dependent, while the quantum states remain fixed.

9.3.1 The Schrödinger picture

The quantum mechanics as we have described it until now, is in fact the Schrödinger picture. The Hilbert space describing the quantum states is spanned by a basis of time-independent ket vectors. We postulated that the quantum state $|\Psi(t)\rangle$ evolves in time according to the Schrödinger equation, maintaining its norm

$$\langle\Psi(t)|\Psi(t)\rangle = 1 \; , \tag{9.21}$$

at any time t.

We know that a unitary transformation satisfies $\hat{U}\hat{U}^\dagger = \hat{U}^\dagger\hat{U} = 1$; i.e., its Hermite conjugate is its inverse. Thus, when applied on a ket $|\Psi\rangle$, a unitary transformation preserves its norm. Indeed, if $|\Psi'\rangle = \hat{U}\,|\Psi\rangle$, then we have

$$\langle\Psi'|\Psi'\rangle = \langle\Psi|\hat{U}^\dagger\hat{U}|\Psi\rangle = \langle\Psi|\Psi\rangle \; . \tag{9.22}$$

Equations (9.21) and (9.22) suggest that the time evolution of quantum states can be described with the help of a unitary transformation $\hat{U}(t,t_0)$, such that

$$|\Psi(t)\rangle = \hat{U}(t,t_0)\,|\Psi(t_0)\rangle \; , \tag{9.23}$$

for all t_0 and t as illustrated in Fig. 9.2.

At this point, we can show that the postulate of unitary time evolution is sufficient to derive Schrödinger equation. We begin by enumerating some of the properties of the time evolution transformation $\hat{U}(t,t_0)$.

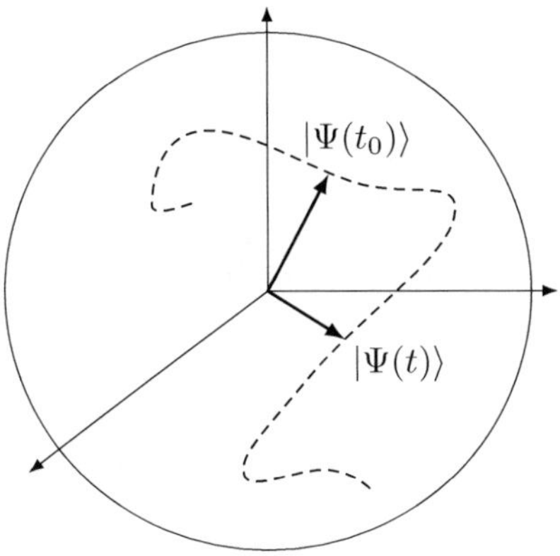

FIGURE 9.2
The state of the system, pictured as a vector in the Hilbert space, evolves by a unitary transformation, such that its norm is preserved.

(1) *Uniqueness*: A system has a unique $\hat{U}$ describing its evolution. Indeed, if there were another operator $\hat{U}'(t, t_0)$ that would generate the same time evolution of $|\Psi(t_0)\rangle$ as $\hat{U}(t, t_0)$ for all t_0 and t, then

$$\hat{U}'(t, t_0) |\Psi(t_0)\rangle = \hat{U}(t, t_0) |\Psi(t_0)\rangle$$

or

$$\left(\hat{U}'(t, t_0) - \hat{U}(t, t_0)\right) |\Psi(t_0)\rangle = 0$$

which implies that $\hat{U}'(t, t_0) = \hat{U}(t, t_0)$ because $|\Psi(t_0)\rangle$ was chosen arbitrarily.

(2) *Identity*: If we set $t = t_0$ in Eq. (9.23), the time evolution gives

$$|\Psi(t_0)\rangle = \hat{U}(t_0, t_0) |\Psi(t_0)\rangle \ .$$

Therefore,

$$\hat{U}(t_0, t_0) = 1 \ . \tag{9.24}$$

(3) *Multiplication rule*: For any two points in time t_0 and t_1, there exists another point t such that the evolution of the system from t_0 to t_1 can be written as the product of two evolution operators, one from t_0 to t and the other from t to t_1. Indeed, we see that

$$|\Psi(t_1)\rangle = \hat{U}(t_1, t) |\Psi(t)\rangle = \hat{U}(t_1, t) \hat{U}(t, t_0) |\Psi(t_0)\rangle \ ,$$

and thus

$$\hat{U}(t_1, t_0) = \hat{U}(t_1, t) \hat{U}(t, t_0) \ . \tag{9.25}$$

(4) *Inverse*: If we choose $t_1 = t_0$ in the multiplication rule, we obtain using the identity property from point (2) above

$$1 = \hat{U}(t_0, t_0) = \hat{U}(t_0, t)\, \hat{U}(t, t_0) \; . \tag{9.26}$$

That is, the inverse of $\hat{U}(t, t_0)$ is $\hat{U}(t_0, t)$, obtained by reversing the two arguments. Because $\hat{U}$ is unitary, we have

$$\hat{U}^{-1}(t, t_0) = \hat{U}^{\dagger}(t, t_0) = \hat{U}(t_0, t) \; . \tag{9.27}$$

To derive Schrödinger equation, we take the time derivative of (9.23). We obtain

$$\frac{\partial}{\partial t} |\Psi(t)\rangle = \frac{\partial \hat{U}(t, t_0)}{\partial t} |\Psi(t_0)\rangle = \frac{\partial \hat{U}(t, t_0)}{\partial t} \hat{U}(t_0, t)\, |\Psi(t)\rangle \; , \tag{9.28}$$

where we used the inverse property (4) to write $|\Psi(t_0)\rangle$ in terms of $|\Psi(t)\rangle$. We now introduce a new operator

$$\hat{Q}(t, t_0) = \frac{\partial \hat{U}(t, t_0)}{\partial t} \hat{U}(t_0, t) \; . \tag{9.29}$$

Then, Eq. (9.28) can be written as

$$\frac{\partial}{\partial t} |\Psi(t)\rangle = \hat{Q}(t, t_0)\, |\Psi(t)\rangle \; . \tag{9.30}$$

The newly introduced operator $\hat{Q}(t, t_0)$ has dimensions of inverse of time, and it has two important properties: it is anti-Hermitian[1], and it doesn't depend on t_0.

To prove the anti-Hermiticity of $\hat{Q}$, we write Eq. (9.26) equivalently as

$$\hat{U}(t, t_0)\, \hat{U}(t_0, t) = 1 \; , \tag{9.31}$$

and take its time derivative. We obtain

$$\frac{\partial \hat{U}(t, t_0)}{\partial t} \hat{U}(t_0, t) + \hat{U}(t, t_0) \frac{\partial \hat{U}(t_0, t)}{\partial t} = 0 \; . \tag{9.32}$$

The first term in Eq. (9.32) is $\hat{Q}(t, t_0)$ and the second is $\hat{Q}^{\dagger}(t, t_0)$, because

$$\hat{Q}^{\dagger}(t, t_0) = \left(\frac{\partial \hat{U}(t, t_0)}{\partial t} \hat{U}(t_0, t) \right)^{\dagger} = \hat{U}^{\dagger}(t_0, t) \frac{\partial \hat{U}^{\dagger}(t, t_0)}{\partial t} = \hat{U}(t, t_0) \frac{\partial \hat{U}(t_0, t)}{\partial t}$$

where we used Eq. (9.27) in the last step. Thus, we have

$$\hat{Q}(t, t_0) + \hat{Q}^{\dagger}(t, t_0) = 0 \; . \tag{9.33}$$

[1] An anti-Hermitian operator $\hat{Q}$ satisfies $\hat{Q}^{\dagger} = -\hat{Q}$.

To prove that $\hat{Q}(t,t_0)$ doesn't depend on t_0, we will show that $\hat{Q}(t,t_0) = \hat{Q}(t,t_1)$ for an arbitrary $t_1 \neq t_0$. Introducing strategically the identity operator $1 = \hat{U}(t_0,t_1)\,\hat{U}(t_1,t_0)$, we have successively:

$$\begin{aligned}
\hat{Q}(t,t_0) &= \frac{\partial \hat{U}(t,t_0)}{\partial t}\hat{U}(t_0,t) = \frac{\partial \hat{U}(t,t_0)}{\partial t}\,1\,\hat{U}(t_0,t) \\
&= \frac{\partial \hat{U}(t,t_0)}{\partial t}\left(\hat{U}(t_0,t_1)\,\hat{U}(t_1,t_0)\right)\hat{U}(t_0,t) \\
&= \frac{\partial\left(\hat{U}(t,t_0)\,\hat{U}(t_0,t_1)\right)}{\partial t}\left(\hat{U}(t_1,t_0)\,\hat{U}(t_0,t)\right) = \frac{\partial \hat{U}(t,t_1)}{\partial t}\hat{U}(t_1,t) \\
&= \hat{Q}(t,t_1)\ .
\end{aligned}$$

It follows that we can drop t_0 from $\hat{Q}$ and write it as $\hat{Q}(t)$. Then, Eq. (9.30) becomes

$$\frac{\partial}{\partial t}\left|\Psi(t)\right\rangle = \hat{Q}(t)\left|\Psi(t)\right\rangle\ , \tag{9.34}$$

where as we have already seen, $\hat{Q}(t)$ is an anti-Hermitian operator. Physical quantities are described by Hermitian operators, so we will multiply Eq. (9.34) by $i\hbar$ and obtain

$$i\hbar\frac{\partial}{\partial t}\left|\Psi(t)\right\rangle = i\hbar\,\hat{Q}(t)\left|\Psi(t)\right\rangle\ . \tag{9.35}$$

We are now in position to define a new operator

$$\hat{H}(t) = i\hbar\,\hat{Q}(t) = i\hbar\,\frac{\partial \hat{U}(t,t_0)}{\partial t}\,\hat{U}(t_0,t)\ , \tag{9.36}$$

which is Hermitian, and has dimensions of energy. $\hat{H}(t)$ is the Hamiltonian of the system. We have thus obtained Schrödinger equation

$$i\hbar\frac{\partial}{\partial t}\left|\Psi(t)\right\rangle = \hat{H}(t)\left|\Psi(t)\right\rangle\ , \tag{9.37}$$

from the postulate of the existence of a time evolution operator.

9.3.1.1 Finding the Time Evolution Operator

In many situations we know the Hamiltonian of the system, and would like to find the time evolution operator. Formally, that can be determined by solving the differential equation

$$i\hbar\,\frac{\partial \hat{U}(t,t_0)}{\partial t} = \hat{H}(t)\,\hat{U}(t,t_0)\ , \tag{9.38}$$

which is obtained from multiplying Eq. (9.36) from the right by $\hat{U}(t,t_0)$.

We first consider an infinitesimal time translation of a quantum state from $|\Psi(t)\rangle$ to $|\Psi(t+\delta t)\rangle$ described by the infinitesimal time evolution operator $\hat{U}(t+\delta t, t)$:

$$\hat{U}(t+\delta t,t)\,|\Psi(t)\rangle = |\Psi(t+\delta t)\rangle = |\Psi(t)\rangle + \frac{\partial}{\partial t}\,|\Psi(t)\rangle\,\delta t\;, \tag{9.39}$$

where, in the Taylor expansion, we ignored terms of order δt^2 and higher. This, can be further written using Schrödinger equation (9.37)

$$\hat{U}(t+\delta t,t)\,|\Psi(t)\rangle = \Big(1 - \frac{i}{\hbar}\hat{H}(t)\,\delta t\Big)\,|\Psi(t)\rangle\;. \tag{9.40}$$

Therefore,

$$\hat{U}(t+\delta t,t) = 1 - \frac{i}{\hbar}\hat{H}(t)\,\delta t\;. \tag{9.41}$$

Because the Hamiltonian $\hat{H}(t)$ appears directly involved in the time translation of the quantum state $|\Psi(t)\rangle$, the Hamiltonian is also known as the *generator* of the time translation.

Next, we will determine the time evolution operator for finite time intervals in two particular cases:

(1) $\hat{H}$ *is time independent*:
Here we will consider the case of a time-independent Hamiltonian, and a finite time translation from t_0 to t. If we divide the time interval $t - t_0$ into n intervals, and let that number increase to ∞, then, the infinitesimal time interval can be seen as the limit $\delta t = \lim_{n\to\infty}(t-t_0)/n$. Then, the time evolution operator yielding a finite time translation will become the infinite product of infinitesimal transformations as given in Eq. (9.41):

$$\hat{U}(t,t_0) = \Big(1 - \frac{i}{\hbar}\hat{H}\,\frac{t-t_0}{n}\Big)\cdots\Big(1 - \frac{i}{\hbar}\hat{H}\,\frac{t-t_0}{n}\Big) = \lim_{n\to\infty}\Big(1 - \frac{i}{\hbar}\hat{H}\,\frac{t-t_0}{n}\Big)^n\,,$$

or

$$\hat{U}(t,t_0) = \exp\big[-\frac{i}{\hbar}\hat{H}\,(t-t_0)\big]\,, \tag{9.42}$$

where the exponential of an operator is understood as the short-hand notation for its series expansion $\exp(\hat{A}) = 1 + \frac{1}{1!}\hat{A} + \frac{1}{2!}\hat{A}^2 + \cdots$.

(2) $\hat{H}$ *depends on time and* $[\hat{H}(t_1), \hat{H}(t_2)] = 0$ *for any* t_1, t_2:
In such a case, we will first divide the time interval $t - t_0$ in sufficiently small intervals, such that the Hamiltonian can be considered constant on each such interval. The time evolution from τ_i to τ_{i+1} will be then given by Eq. (9.42), and the overall time evolution will be

$$\begin{aligned}
\hat{U}(t,t_0) &= \hat{U}(t,\tau_1)\hat{U}(\tau_1,\tau_2)\cdots\hat{U}(\tau_{n-2},\tau_{n-1})\hat{U}(\tau_{n-1},t_0)\\
&= e^{-\frac{i}{\hbar}\hat{H}(\tau_1)\,(t-\tau_1)}e^{-\frac{i}{\hbar}\hat{H}(\tau_2)\,(\tau_1-\tau_2)}\cdots e^{-\frac{i}{\hbar}\hat{H}(t_0)\,(\tau_{n-1}-t_0)}\\
&= \exp\left[-\frac{i}{\hbar}\sum_{i=1}^{n}\hat{H}(\tau_i)\,\Delta t_i\right]\,,\quad \text{where } \Delta t_i = \tau_i - \tau_{i+1}\,.
\end{aligned}$$

In the last line we could write the product of the exponential factors as an exponential of the sum because the operators involved, $\hat{H}(\tau_i)$, commute among themselves (see also Problem 9.1). This sum is exactly the Riemann sum. Thus, in the limit $\Delta t_i \to 0$, it becomes the definite integral

$$\hat{U}(t, t_0) = \exp\left[-\frac{i}{\hbar}\int_{t_0}^{t} \hat{H}(t')\, dt'\right] . \tag{9.43}$$

The general case when $[\hat{H}(t_1), \hat{H}(t_2)] \neq 0$ will be addressed later in section 11.2.1.

If we have an observable $\hat{A}$ that doesn't depend explicitly on time, then its expectation value will evolve in time according to

$$\frac{d}{dt}\langle\Psi(t)|\hat{A}|\Psi(t)\rangle = \frac{i}{\hbar}\langle\Psi(t)|[\hat{H}, \hat{A}]|\Psi(t)\rangle \tag{9.44}$$

as it can be shown with the help of Schrödinger equation (see also Problem 9.2).

A time independent operator $\hat{A}$ is said to be *conserved* if it commutes with the Hamiltonian. From Eq.(9.44) it follows that for a conserved operator, the expectation value $\langle A\rangle$ is constant in time.

9.3.2 The Heisenberg picture

In the Schrödinger picture we described the dynamics of systems using the time evolution of quantum states. In the Heisenberg picture we confine the dynamics to the operators, while keeping the quantum states fixed in time. To this end, we define time-dependent Heisenberg operators starting from Schrödinger operators. Schrödinger operators are those we have been regularly using so far such as $\hat{x}$ and $\hat{p}$, as well as the operators constructed from them, possibly depending on time, like Hamiltonians describing time dependent interactions.

In the Heisenberg picture, quantum states $|\Psi_H\rangle$ are defined from Schrödinger states $|\Psi_S(t)\rangle$ via

$$|\Psi_H\rangle = \hat{U}^{-1}(t, 0)\,|\Psi_S(t)\rangle = \hat{U}^{\dagger}(t, 0)\,|\Psi_S(t)\rangle = |\Psi_S(0)\rangle \ , \tag{9.45}$$

so that they are constant in time. What about the operators? We consider a Schrödinger operator $\hat{\Lambda}_S$, where we added the subscript S for Schrödinger, and examine the matrix element of $\hat{\Lambda}_S$ in between two time-dependent states $|\Psi_S(t)\rangle$ and $|\Phi_S(t)\rangle$

$$\begin{aligned}\langle\Phi_S(t)|\hat{\Lambda}_S|\Psi_S(t)\rangle &= \langle\Phi_S(0)|\hat{U}^{\dagger}(t, 0)\,\hat{\Lambda}_S\,\hat{U}(t, 0)|\Psi_S(0)\rangle \\ &= \langle\Phi_H|\hat{U}^{\dagger}(t, 0)\,\hat{\Lambda}_S\,\hat{U}(t, 0)|\Psi_H\rangle \ .\end{aligned} \tag{9.46}$$

This suggests that we define the operator $\hat{\Lambda}_H$ in the Heisenberg picture by

$$\hat{\Lambda}_H(t) = \hat{U}^{\dagger}(t, 0)\,\hat{\Lambda}_S\,\hat{U}(t, 0) \ . \tag{9.47}$$

Consequently,

$$\langle \Phi_S(t)|\hat{\Lambda}_S|\Psi_S(t)\rangle = \langle \Phi_H|\hat{\Lambda}_H(t)|\Psi_H\rangle \ . \tag{9.48}$$

There are several properties of Heisenberg operators which are important:

(1) At $t = 0$, an Heisenberg operator becomes identical to its Schrödinger counterpart

$$\hat{\Lambda}_H(0) = \hat{\Lambda}_S \ .$$

(2) Operators satisfy the same commutation relations in Heisenberg picture as in the Schrödinger picture. Indeed, if $[\hat{A}_S, \hat{B}_S] = \hat{C}_S$, then

$$\begin{aligned}[\hat{A}_H, \hat{B}_H] &= \hat{A}_H\hat{B}_H - \hat{B}_H\hat{A}_H = \hat{U}^\dagger(\hat{A}_S\hat{U}\hat{U}^\dagger\hat{B}_S - \hat{B}_S\hat{U}\hat{U}^\dagger\hat{A}_S)\hat{U} \\ &= \hat{U}^\dagger([A_S, B_S])\hat{U} = \hat{U}^\dagger\hat{C}_S\hat{U} \\ &= \hat{C}_H \ .\end{aligned}$$

For example, using the fact that the identity operator doesn't change with the picture, $1_H = 1_S \equiv 1$, we have

$$[\hat{x}, \hat{p}] = i\hbar \Rightarrow [\hat{x}_H(t), \hat{p}_H(t)] = i\hbar \ .$$

(3) In the Schrödinger picture the operators $\hat{x}$ and $\hat{p}$ appear in products inside the Hamiltonian. It follows that, in the Heisenberg picture, the Hamiltonian takes the same form as in the Schrödinger picture, but with $\hat{x}$ and $\hat{p}$ replaced by $\hat{x}_H(t)$ and $\hat{p}_H(t)$, respectively.

9.3.2.1 Heisenberg equations of motion

In the Heisenberg picture, operators satisfy a differential equation, the Heisenberg equation of motion . To find this equation, we multiply Eq. (9.47) by $i\hbar$ and take its time derivative. We obtain

$$\begin{aligned} i\hbar\frac{d}{dt}\hat{\Lambda}_H &= \Big(i\hbar\,\frac{\partial \hat{U}^\dagger(t,0)}{\partial t}\Big)\hat{\Lambda}_S\,\hat{U}(t,0) + \hat{U}^\dagger(t,0)\Big(i\hbar\,\frac{\partial\hat{\Lambda}_S}{\partial t}\Big)\hat{U}(t,0) \\ &+ \hat{U}^\dagger(t,0)\,\hat{\Lambda}_S\Big(i\hbar\,\frac{\partial \hat{U}(t,0)}{\partial t}\Big) \\ &= -\hat{U}(t,0)^\dagger\hat{H}(t)\,\hat{\Lambda}_S\hat{U}(t,0) + \hat{U}^\dagger(t,0)\Big(i\hbar\,\frac{\partial\hat{\Lambda}_S}{\partial t}\Big)\hat{U}(t,0) \\ &+ \hat{U}^\dagger(t,0)\,\hat{\Lambda}_S\hat{H}(t)\,\hat{U}(t,0) \\ &= \hat{\Lambda}_H(t)\hat{H}_H(t) - \hat{H}_H(t)\hat{\Lambda}_H(t) + i\hbar\Big(\frac{\partial\hat{\Lambda}_S}{\partial t}\Big)_H , \end{aligned} \tag{9.49}$$

where we made use of Eq. (9.38), its Hermite conjugate, and the fact that $\hat{U}^\dagger\hat{U} = 1$. Because the operations of first taking the time derivative and then going into the Heisenberg picture commute, we obtain finally

$$i\hbar\frac{d}{dt}\hat{\Lambda}_H = [\hat{\Lambda}_H(t), \hat{H}_H(t)] + i\hbar\Big(\frac{\partial\hat{\Lambda}_H}{\partial t}\Big) \ . \tag{9.50}$$

If the operator $\hat{\Lambda}_S$ doesn't depend on time, then the last term in Eq. (9.50) vanishes, and the Heisenberg equation becomes

$$i\hbar\frac{d}{dt}\hat{\Lambda}_H = [\hat{\Lambda}_H(t), \hat{H}_H(t)] \ . \tag{9.51}$$

Example 9.1 Harmonic oscillator in Heisenberg picture

We want to find the explicit form of the Heisenberg operators for the harmonic oscillator.

We start with the Schrödinger Hamiltonian

$$\hat{H}_S = \frac{\hat{p}^2}{2m} + \frac{1}{2}m\omega^2\hat{x}^2 \ ,$$

and its corresponding Heisenberg operator

$$\hat{H}_H(t) = \frac{\hat{p}_H^2(t)}{2m} + \frac{1}{2}m\omega^2\hat{x}_H^2(t) \ .$$

Because $\hat{x}$ and $\hat{p}$ do not depend explicitly on time, we will use Heisenberg equation of motion (9.51). We have

$$\begin{aligned} i\hbar\frac{d}{dt}\hat{x}_H(t) &= [\hat{x}_H(t), \hat{H}_H(t)] \\ &= \frac{1}{2m}[\hat{x}_H(t), \hat{p}_H^2(t)] = \frac{2\hat{p}_H(t)}{2m}[\hat{x}_H(t), \hat{p}_H(t)] \\ &= i\hbar\frac{\hat{p}_H(t)}{m} \ , \end{aligned}$$

where in the last line we used the fact that the commutation relations in the Heisenberg picture are the same as in the Schrödinger picture. We thus obtain the Heisenberg equation of motion for the position operator

$$\frac{d}{dt}\hat{x}_H(t) = \frac{\hat{p}_H(t)}{m} \ , \tag{9.52}$$

which looks very much like its classical counterpart. The momentum equation of motion in Heisenberg picture gives

$$\begin{aligned} i\hbar\frac{d}{dt}\hat{p}_H(t) &= [\hat{p}_H(t), \hat{H}_H(t)] \\ &= \frac{m\omega^2}{2}[\hat{p}_H(t), \hat{x}_H^2(t)] = \frac{2m\omega^2\hat{x}_H(t)}{2}[\hat{p}_H(t), \hat{x}_H(t)] \\ &= -i\hbar m\omega^2\hat{x}_H(t) \ . \end{aligned}$$

Therefore,

$$\frac{d}{dt}\hat{p}_H(t) = -m\omega^2\hat{x}_H(t) \ . \tag{9.53}$$

To find the explicit time dependence of $\hat{x}_H(t)$ and $\hat{p}_H(t)$ we take the time derivative of Eq. (9.52) and combine it with (9.53). We obtain

$$\frac{d^2}{dt^2}\hat{x}_H(t) = -\omega^2 \hat{x}_H(t) , \tag{9.54}$$

whose solution is

$$\hat{x}_H(t) = \hat{C}_1 \cos\omega t + \hat{C}_2 \sin\omega t , \tag{9.55}$$

where $\hat{C}_1$ and $\hat{C}_2$ are constant operators to be determined from the initial conditions. Substituting now Eq. (9.55) into (9.52) we get

$$\hat{p}_H(t) = m\frac{d}{dt}\hat{x}_H(t) = -m\omega\, \hat{C}_1 \sin\omega t + m\omega\, \hat{C}_2 \cos\omega t . \tag{9.56}$$

To find the values of the time-independent operators $\hat{C}_1$ and $\hat{C}_2$ we use the fact that at $t = 0$ the Heisenberg operators and the corresponding Schrödinger operators coincide. Thus, from Eqs. (9.55) and(9.56), we obtain

$$\hat{x}_H(0) = \hat{x} = \hat{C}_1 ; \qquad \hat{p}_H(0) = \hat{p} = m\omega\, \hat{C}_2 .$$

Therefore, the equations of motion of the Heisenberg operators in the case of the harmonic oscillator are:

$$\hat{x}_H(t) = \hat{x}\cos\omega t + \frac{1}{m\omega}\,\hat{p}\sin\omega t \tag{9.57}$$

$$\hat{p}_H(t) = -m\omega\,\hat{x}\sin\omega t + \hat{p}\cos\omega t . \tag{9.58}$$

9.4 Problems

9.1 The exponential of an operator $\hat{K}$ is defined by the formal Taylor expansion $e^{\hat{K}} = 1 + \hat{K} + \frac{1}{2!}\hat{K}^2 + \cdots$. Show that for two commuting operators, $\hat{A}\hat{B} = \hat{B}\hat{A}$, we have

$$e^{\hat{A}}e^{\hat{B}} = e^{\hat{A}+\hat{B}} .$$

Note: This is a particular case of a general result known as the Baker-Campbell-Hausdorff theorem.

9.2 The time derivative of the expectation value of an observable $\hat{A}$ in a state $|\Psi\rangle$ can be written as

$$\frac{d}{dt}\langle A\rangle = \frac{d}{dt}\langle\Psi|\hat{A}|\Psi\rangle = \Big(\frac{\partial}{\partial t}\langle\Psi|\Big)\hat{A}\,|\Psi\rangle + \langle\Psi|\Big(\frac{\partial}{\partial t}\hat{A}\Big)|\Psi\rangle + \langle\Psi|\,\hat{A}\Big(\frac{\partial}{\partial t}|\Psi\rangle\Big) .$$

Use Schrödinger equation together with its conjugate, to show that

$$\frac{d}{dt}\langle A\rangle = \frac{i}{\hbar}\langle[\hat{H},\hat{A}]\rangle + \langle\partial A/\partial t\rangle .$$

10

Time Independent Perturbation Theory

Many physical systems cannot be solved exactly. In this case we look for approximate methods to gain insight into their properties. Perturbation theory is one such method. When the Hamiltonian of the system doesn't change in time, one approximate method is the Time Independent Perturbation Theory.

10.1 Introduction

The infinite square well, the harmonic oscillator, and the hydrogen atom are among the few physical systems for which the Schrödinger equation can be solved exactly. For this reason they are described in detail in most of the introductory quantum mechanics textbooks. However, in practice, most physical systems are not exactly solvable, i.e., they are described by Schrödinger equations whose solutions cannot be expressed in terms of commonly used functions. In order to study such systems we try to decompose the corresponding Hamiltonian H into a solvable Hamiltonian H_0 plus a "small" perturbation V_1. For such cases we say that the Hamiltonian is "close enough" to an exactly solvable Hamiltonian to warrant the use of perturbative methods to approximate eigenenergies and eigenfunctions.

10.2 Perturbation theory: the non-degenerate case

To illustrate the perturbation theory, we start with the simple case of non-degenerate systems. We will consider a system whose Hamiltonian $\hat{H}$ can be written as

$$\hat{H} = \hat{H}_0 + \hat{V}_1 \ , \tag{10.1}$$

where $\hat{H}_0$ is the Hamiltonian of an exactly solvable system and $\hat{V}_1$ is the perturbation, considered small. The perturbation method begins with the eigenvalue problem for $\hat{H}_0$

$$\hat{H}_0 \, \phi_n^{(0)} = E_n^{(0)} \, \phi_n^{(0)} \ , \tag{10.2}$$

DOI: 10.1201/9781003174967-10

where $E_n^{(0)}$ is the eigenenergy corresponding to the eigenfunction $\phi_n^{(0)}$, both known exactly. We assume that $\phi_n^{(0)}$ are normalized. The goal is to find approximately the eigenenergies and eigenfunction of $\hat{H}$:

$$\hat{H}\,\psi_n = E_n\,\psi_n \,. \tag{10.3}$$

There are several steps in the process. We first introduce a dimensionless parameter λ, where $0 \leq \lambda \leq 1$, and write

$$\begin{aligned} \hat{H} &= \hat{H}_0 + \lambda \hat{V}_1 \\ E_n &= E_n^{(0)} + \lambda E_n^{(1)} + \lambda^2 E_n^{(2)} + \cdots \\ \psi_n &= \phi_n^{(0)} + \lambda \phi_n^{(1)} + \lambda^2 \phi_n^{(2)} + \cdots \end{aligned} \tag{10.4}$$

by formally expanding both the eigenenergies and eigenfunctions in powers of λ. The dimensionless parameter λ varies from 0 —no perturbation, to 1 —full perturbation. We substitute the series expansions of E_n and ψ_n in Eq. (10.3) and then equate the terms corresponding to the same powers of λ on both sides of the equation. Thus, Eq. (10.3) gives

$$\begin{aligned} &(\hat{H}_0 + \lambda \hat{V}_1)(\phi_n^{(0)} + \lambda \phi_n^{(1)} + \lambda^2 \phi_n^{(2)} + \cdots) \\ &\quad = (E_n^{(0)} + \lambda E_n^{(1)} + \lambda^2 E_n^{(2)} + \cdots)(\phi_n^{(0)} + \lambda \phi_n^{(1)} + \lambda^2 \phi_n^{(2)} + \cdots) \,. \end{aligned} \tag{10.5}$$

We now require that the coefficients of the same powers of λ on both sides of this equation be equal. The first three orders are

- Zero order in λ:

$$\hat{H}_0\,\phi_n^{(0)} = E_n^{(0)}\,\phi_n^{(0)} \,, \tag{10.6}$$

- First order in λ:

$$\hat{V}_1\,\phi_n^{(0)} + \hat{H}_0\,\phi_n^{(1)} = E_n^{(1)}\,\phi_n^{(0)} + E_n^{(0)}\,\phi_n^{(1)} \,, \tag{10.7}$$

- Second order in λ:

$$\hat{V}_1\,\phi_n^{(1)} + \hat{H}_0\,\phi_n^{(2)} = E_n^{(2)}\,\phi_n^{(0)} + E_n^{(1)}\,\phi_n^{(1)} + E_n^{(0)}\,\phi_n^{(2)} \,. \tag{10.8}$$

Exercise 10.1 Derive Eqs. (10.6) – (10.8).

Note that Eq. (10.6) is just the unperturbed problem, which we assume is known.

To find $E_n^{(1)}$, the first order correction to the energy, we take the inner product of Eq. (10.7) with $\phi_n^{(0)}$:

$$\langle \phi_n^{(0)} | \hat{V}_1 | \phi_n^{(0)} \rangle + \langle \phi_n^{(0)} | \hat{H}_0 | \phi_n^{(1)} \rangle = E_n^{(1)}\,\langle \phi_n^{(0)} | \phi_n^{(0)} \rangle + E_n^{(0)}\,\langle \phi_n^{(0)} | \phi_n^{(1)} \rangle \,. \tag{10.9}$$

Because $\hat{H}_0$ is Hermitian, we have $\langle\phi_n^{(0)}|\hat{H}_0|\phi_n^{(1)}\rangle = E_n^{(0)}\langle\phi_n^{(0)}|\phi_n^{(1)}\rangle$, and thus, Eq. (10.9) yields the *first order correction* for the eigenenergy

$$E_n^{(1)} = \langle\phi_n^{(0)}|\hat{V}_1|\phi_n^{(0)}\rangle \equiv V_{nn} \;, \tag{10.10}$$

where we introduced the notation $V_{mn} = \langle\phi_m^{(0)}|\hat{V}_1|\phi_n^{(0)}\rangle$.

Before we go any further, it is worth pausing to reflect on the higher order terms in the expansion of ψ_n from Eq. (10.4). We start with the first order term by rewriting Eq. (10.7) as

$$(\hat{H}_0 - E_n^{(0)})\,\phi_n^{(1)} = -(\hat{V}_1 - E_n^{(1)})\,\phi_n^{(0)} \;, \tag{10.11}$$

where now, in the right-hand side, all quantities are known. Because the set $\phi_n^{(0)}$ is complete, we can express $\phi_n^{(1)}$ as a linear combination

$$\phi_n^{(1)} = \sum_{k\neq n} C_n^k \phi_k^{(0)} \;, \tag{10.12}$$

where C_n^k are appropriately chosen constants. If $\phi_n^{(1)}$ would have any component along $\phi_n^{(0)}$, say $\alpha\,\phi_n^{(0)}$, that would not change Eq. (10.11) because the corresponding term in the left-hand side would vanish. Therefore, for the first order in λ, we can choose $\alpha = 0$, and thus $\langle\phi_n^{(0)}|\phi_n^{(1)}\rangle = 0$. Thus, we impose that $\psi = \phi_n^{(0)} + \phi_n^{(1)}$ is normalized up to, and including terms of the first order[28]. Similarly, for the higher order terms.

To find the first order correction to the wave function, we will calculate the coefficients C_n^k. To this end, we substitute Eq. (10.12) back into Eq. (10.7) and then take the inner product with $\phi_j^{(0)}$. This yields

$$\langle\phi_j^{(0)}|\hat{V}_1|\phi_n^{(0)}\rangle + \sum_{k\neq n} C_n^k E_j^{(0)}\langle\phi_j^{(0)}|\phi_k^{(0)}\rangle = E_n^{(1)}\langle\phi_j^{(0)}|\phi_n^{(0)}\rangle + \sum_{k\neq n} C_n^k E_n^{(0)}\langle\phi_j^{(0)}|\phi_k^{(0)}\rangle \tag{10.13}$$

If $j = n$, we recover Eq. (10.10). If $j \neq n$, then using the orthonormality of $\phi_n^{(0)}$, we obtain that only the term $k = j$ survives in both sums of Eq. (10.13). Thus, we get

$$C_n^j = \frac{\langle\phi_j^{(0)}|\hat{V}_1|\phi_n^{(0)}\rangle}{E_n^{(0)} - E_j^{(0)}} \;. \tag{10.14}$$

Therefore, the first order correction to the eigenfunction ψ_n is

$$\phi_n^{(1)} = \sum_{k\neq n} \frac{\langle\phi_k^{(0)}|\hat{V}_1|\phi_n^{(0)}\rangle}{E_n^{(0)} - E_k^{(0)}}\,\phi_k^{(0)} \equiv \sum_{k\neq n} \frac{V_{kn}}{\Delta_{nk}}\,\phi_k^{(0)} \tag{10.15}$$

where we denoted $\Delta_{nk} = E_n^{(0)} - E_k^{(0)}$.

To obtain the second order correction in the energy, we take the inner product of Eq. (10.8) with $\phi_n^{(0)}$. We get

$$\langle\phi_n^{(0)}|\hat{V}_1|\phi_n^{(1)}\rangle + \langle\phi_n^{(0)}|\hat{H}_0\,|\phi_n^{(2)}\rangle = E_n^{(2)}\,\langle\phi_n^{(0)}|\phi_n^{(0)}\rangle + E_n^{(1)}\,\langle\phi_n^{(0)}|\phi_n^{(1)}\rangle + E_n^{(0)}\,\langle\phi_n^{(0)}|\phi_n^{(2)}\rangle$$

As before, the hermiticity of $\hat{H}_0$ gives $\langle\phi_n^{(0)}|\hat{H}_0\,|\phi_n^{(2)}\rangle = E_n^{(0)}\,\langle\phi_n^{(0)}|\phi_n^{(2)}\rangle$, which together with $\langle\phi_n^{(0)}|\phi_n^{(1)}\rangle = 0$, $\langle\phi_n^{(0)}|\phi_n^{(0)}\rangle = 1$, simplifies the above equation to

$$E_n^{(2)} = \langle\phi_n^{(0)}|\hat{V}_1|\phi_n^{(1)}\rangle = \sum_{k\neq n} \frac{\langle\phi_n^{(0)}|\hat{V}_1|\phi_k^{(0)}\rangle\,\langle\phi_k^{(0)}|\hat{V}_1|\phi_n^{(0)}\rangle}{E_n^{(0)} - E_k^{(0)}} = \sum_{k\neq n} \frac{|\,\langle\phi_n^{(0)}|\hat{V}_1|\phi_k^{(0)}\rangle\,|^2}{E_n^{(0)} - E_k^{(0)}}\ ,$$

where we used the known form of $\phi_n^{(1)}$ from Eq. (10.15). We obtain finally for the second order correction in the energy

$$E_n^{(2)} = \sum_{k\neq n} \frac{|V_{nk}|^2}{\Delta_{nk}}\ . \tag{10.16}$$

This process can be iterated further. Knowing $E_n^{(2)}$ we can then find $\phi_n^{(2)}$, which then allows us to find $E_n^{(3)}$, and so on. For example, the third order correction in energy is given by [28]

$$E_n^{(3)} = \sum_{l,k\neq n} \frac{V_{nl}V_{lk}V_{kn}}{\Delta_{nl}\Delta_{nk}} - V_{nn}\sum_{k\neq n} \frac{|V_{nk}|^2}{\Delta_{nk}^2}\ . \tag{10.17}$$

Note that successive orders depend on the ability to calculate infinite sums, which in general is difficult to achieve[1]. However, in many cases, the first and respectively second order corrections give results close enough to the experimental values.

Example 10.1 Stark effect: perturbing the ground state
Let us consider a hydrogen atom in its ground state $|100\rangle$ subject to a weak electric field directed along the positive z-axis, $\vec{\mathcal{E}} = \mathcal{E}\vec{k}$. We ignore the spin for this example. In general, the eigenstates of the hydrogen atom are degenerate, except for the ground state. Thus, for the ground state, we can deploy the machinery we developed for the non-degenerate perturbation theory. In the next section, we will consider the same example for the first excited state, which is degenerate.

We will now find an approximate value of the polarizability of the hydrogen atom by estimating the energy shift to the second-order in perturbation. Note that the polarizability α of an atom which is subject to an electric field $\vec{\mathcal{E}}$ is given in terms of the energy shift ΔE by $\alpha = -2\Delta E/\mathcal{E}^2$.

[1]There are other methods, such as the logarithmic perturbation theory, which often allows for easier calculation of these higher orders.

We start by writing the Hamiltonian of the unperturbed hydrogen atom

$$\hat{H}_0 = -\frac{\hbar^2}{2m}\nabla^2 - \frac{e^2}{4\pi\epsilon_0}\frac{1}{r} \; . \tag{10.18}$$

When the electric field is turned on, it generates an interaction potential $\hat{V}_1 = e\vec{\mathcal{E}} \cdot \hat{\vec{r}} = e\mathcal{E}\hat{z} = e\mathcal{E}z$, that is added to $\hat{H}_0$. We consider the case of a weak electric field such that $\hat{V}_1$ is small, and use Eqs. (10.10) and (10.16) from the non-degenerate perturbation theory. We obtain for the energy of the system to second-order perturbation

$$E_{100} = E_{100}^{(0)} + e\mathcal{E}\,\langle 100|\hat{z}|100\rangle + e^2\mathcal{E}^2 \sum_{nlm\neq 100} \frac{|\,\langle nlm|\hat{z}|100\rangle\,|^2}{E_{100}^{(0)} - E_{nlm}^{(0)}} \; . \tag{10.19}$$

The first correction term gives zero

$$E_{100}^{(1)} = e\mathcal{E}\,\langle 100|\hat{z}|100\rangle = e\mathcal{E}\int |\psi_{100}(\vec{r})|^2 z\, d^3r = 0 \; , \tag{10.20}$$

because $|\psi_{100}(\vec{r})|^2$ is even under parity, while z is odd.

This implies that there is no linear Stark effect. The lowest correction is quadratic in the electric field $\mathcal{E}$. We are thus left with only the second order correction in energy

$$E_{100}^{(2)} = e^2\mathcal{E}^2 \sum_{nlm\neq 100} \frac{|\,\langle nlm|\hat{z}|100\rangle\,|^2}{E_{100}^{(0)} - E_{nlm}^{(0)}} \; . \tag{10.21}$$

It is rather difficult to calculate the sum in Eq. (10.21), therefore we are going to make an estimate. Since $n \geq 2$ we have [19]

$$E_{100}^{(0)} - E_{nlm}^{(0)} \leq E_{100}^{(0)} - E_{200}^{(0)} = -\frac{e^2}{2a_0}\frac{1}{4\pi\epsilon_0}\left(1 - \frac{1}{2^2}\right) = -\frac{3e^2}{8a_0}\frac{1}{4\pi\epsilon_0} \tag{10.22}$$

where $a_0 = 4\pi\epsilon_0\hbar^2/me^2 \cong 5.29 \times 10^{-11}$m is the Bohr radius. Thus, Eq. (10.21) combined with 10.20 and 10.19 yields

$$\Delta E \equiv E_{100}^{(2)} \geq -\frac{8a_0}{3}\,(4\pi\epsilon_0)\,\mathcal{E}^2 \sum_{nlm\neq 100} |\,\langle nlm|\hat{z}|100\rangle\,|^2 \; . \tag{10.23}$$

Thus, the polarizability of the hydrogen atom is bounded by

$$\alpha = -\frac{2\Delta E}{\mathcal{E}^2} \leq \frac{16a_0}{3}\,(4\pi\epsilon_0) \sum_{nlm\neq 100} |\,\langle nlm|\hat{z}|100\rangle\,|^2 \; . \tag{10.24}$$

The sum in Eq. (10.24) can be calculated exactly, using the fact that $\langle 100|\hat{z}|100\rangle = 0$ and that the set of states $|nlm\rangle$ is complete [2]. We have

$$\begin{aligned}\sum_{nlm\neq 100} |\langle nlm|\hat{z}|100\rangle|^2 &= \sum_{\text{all } nlm} \langle 100|\hat{z}|nlm\rangle \langle nlm|\hat{z}|100\rangle \\ &= \langle 100|\,\hat{z}\Big(\sum_{\text{all } nlm} |nlm\rangle\langle nlm|\Big)\,\hat{z}\,|100\rangle \\ &= \langle 100|\hat{z}^2|100\rangle \ . \end{aligned} \tag{10.25}$$

Because $\psi_{100}(r,\theta,\varphi) = R_{1,0}(r)Y_{0,0}(\theta,\varphi) = 2a_0^{-3/2}\exp(-r/a_0)(4\pi)^{-1/2}$ and $z = r\cos\theta$, we obtain

$$\langle 100|\hat{z}^2|100\rangle = \frac{1}{4\pi}\int_0^\infty r^4 R_{10}^2(r)\,dr\int_0^\pi \sin\theta\cos^2\theta d\theta\int_0^{2\pi} d\varphi = a_0^2\ , \tag{10.26}$$

which when substituted into Eq. (10.24) gives the desired approximation

$$\alpha \leq \frac{16\,a_0^3}{3}(4\pi\epsilon_0)\ . \tag{10.27}$$

It turns out that the sum in Eq. (10.21) can be calculated exactly [43] as well, which gives the exact value for the polarizability as $\alpha = \frac{9\,a_0^3}{2}(4\pi\epsilon_0)$.

Exercise 10.2 Prove Eqs. (10.26) and (10.27).

The complicated issue of calculating the sums over all the states can be mitigated using a slightly modified technique, known as the *logarithmic perturbation* method.

10.2.1 The logarithmic perturbation theory

The logarithmic perturbation method [23] avoids infinite sums over all possible eigenstates, which often cannot be explicitly performed.

The Schrödinger equation for the perturbed system reads

$$-\frac{\hbar^2}{2m}\frac{d^2\psi(x)}{dx^2} + V(x)\psi(x) = E\psi(x)\ , \tag{10.28}$$

where $V(x) = V_0(x) + V_1(x)$ is the perturbed potential. As before, V_0 is the unperturbed potential for which we know the eigenfunctions $\phi_n^{(0)}(x)$ and the corresponding eigenenergies E_n, while $V_1(x)$ is the perturbation, considered "small." For simplicity, we will drop the subscript n and the superscript (0)

[2]Refer to Eq. (9.11) also.

and write E, ψ, ϕ, to denote $E_n, \psi_n, \phi_n^{(0)}$ for any n. To jump-start the perturbation procedure we introduce the expansion parameter $0 \leq \lambda \leq 1$, and write $V(x) = V_0(x) + \lambda V_1(x)$. Thus, $\lambda = 0$ models the unperturbed, exactly solvable potential, while $\lambda = 1$ gives the fully perturbed potential.

We start with the Schrödinger equation, where we denote by prime the x-derivative

$$-\frac{\hbar^2}{2m}\psi''(x) + [V_0(x) + \lambda V_1(x)]\psi(x) = E\psi(x) \ , \tag{10.29}$$

and define a new function

$$S(x) = \ln \psi(x) \ , \tag{10.30}$$

which converts the Schrödinger equation (10.29) into a Riccati equation

$$S''(x) + S'(x)^2 = \frac{2m}{\hbar^2}\left[V_0(x) + \lambda V_1(x) - E \right] \ . \tag{10.31}$$

Then, we expand E and $S'(x)$ in powers of λ, and obtain

$$E = E^{(0)} + \lambda E^{(1)} + \lambda^2 E^{(2)} + \cdots \ , \tag{10.32}$$

$$S'(x) = R_0(x) + \lambda R_1(x) + \lambda^2 R_2(x) + \cdots \ . \tag{10.33}$$

Substituting Eqs. (10.32) and (10.33) into Eq. (10.31), we obtain for the first three successive powers of λ

- Zero order in λ:

$$R_0'(x) + R_0^2(x) = \frac{2m}{\hbar^2}\left[V_0(x) - E^{(0)} \right] \tag{10.34}$$

- First order in λ:

$$R_1'(x) + 2R_0(x)R_1(x) = \frac{2m}{\hbar^2}\left[V_1(x) - E^{(1)} \right] \tag{10.35}$$

- Second order in λ:

$$R_2'(x) + 2R_0(x)R_2(x) = -\frac{2m}{\hbar^2} E^{(2)} - R_1^2(x) \tag{10.36}$$

In general, for $n \geq 2$ we have

$$R_n'(x) + 2R_0(x)R_n(x) = -\frac{2m}{\hbar^2} E^{(n)} - \sum_{j=1}^{n-1} R_j(x)R_{n-j}(x) \ . \tag{10.37}$$

Exercise 10.3 Derive Eqs. (10.34)–(10.37) and show that Eq. (10.34) is just the unperturbed problem.

At this stage, we see that each of the differential equations can be integrated using $|\phi(x)|^2 = \exp[2\int^x R_0(y)\,dy]$ as an integrating factor. The solution of the first equation in the perturbation expansion Eq. (10.34) is just $\phi(x)$, the unperturbed eigenfunction, which is known.

The first order correction

To calculate the first order correction of the energy we multiply Eq. (10.35) with the integrating factor $|\phi(x)|^2 = \exp[2\int^x R_0(y)\,dy]$. We obtain successively

$$[R_1'(x) + 2R_0(x)R_1(x)]\,|\phi(x)|^2 = \frac{2m}{\hbar^2}\left[V_1(x) - E^{(1)}\right]|\phi(x)|^2$$

$$\frac{d}{dx}\left[R_1(x)|\phi(x)|^2\right] = \frac{2m}{\hbar^2}\left[V_1(x) - E^{(1)}\right]|\phi(x)|^2$$

or

$$R_1(x)|\phi(x)|^2 = \frac{2m}{\hbar^2}\int_{-\infty}^{x}\left[V_1(y) - E^{(1)}\right]|\phi(y)|^2 dy\ , \tag{10.38}$$

where we used the fact that the normalizability of $\phi(x)$ requires that $\phi(x) \to 0$ when $x \to \pm\infty$. Thus, the left-hand side of Eq. (10.38) vanishes in the limit $x \to +\infty$. We obtain:

$$E^{(1)} = \int_{-\infty}^{+\infty} V_1(y)\,|\phi(y)|^2 dy = \langle\phi|V_1|\phi\rangle\ . \tag{10.39}$$

This is the standard first-order correction for energy which we already found in the previous section, Eq. (10.10). Once we found $E^{(1)}$ we can go back into Eq. (10.38) and obtain $R_1(x)$:

$$R_1(x) = \frac{2m}{\hbar^2}|\phi(x)|^{-2}\int_{-\infty}^{x}\left[V_1(y) - E^{(1)}\right]|\phi(y)|^2 dy\ , \tag{10.40}$$

which becomes essential for calculating the next order correction in energy.

The second order correction

For the second order correction, we start with Eq. (10.36) and use the same integrating factor $|\phi(x)|^2$. We obtain

$$\frac{d}{dx}\left[R_2(x)|\phi(x)|^2\right] = \left[-\frac{2m}{\hbar^2}E^{(2)} - R_1^2(x)\right]|\phi(x)|^2\ ,$$

which gives

$$E^{(2)} = -\frac{\hbar^2}{2m}\int_{-\infty}^{+\infty} R_1^2(y)\,|\phi(y)|^2 dy\ . \tag{10.41}$$

From Eq. (10.40) we observe that $R_1(x)$ is a real function, hence the second order correction in energy, Eq. (10.41), is always less than zero. Substituting Eq. (10.40) into Eq. (10.41), we obtain finally

$$E^{(2)} = -\frac{2m}{\hbar^2}\int_{-\infty}^{+\infty} |\phi(x)|^{-2}\left[\int_{-\infty}^{x}\left[V_1(y) - E^{(1)}\right]|\phi(y)|^2 dy\right]^2 dx \, . \quad (10.42)$$

This formula is, in general, easier to compute than the sum over all possible eigenfunctions as prescribed by Eq. (10.16).

Example 10.2 Harmonic oscillator with perturbation $\mathbf{V_1(x)} = \epsilon \mathbf{x} \mathbf{e}^{-\beta \mathbf{x}^2}$
As an example of the logarithmic perturbation method, we will compute the case of the harmonic oscillator with the perturbing potential $V_1(x) = \epsilon x e^{-\beta x^2}$.

The unperturbed ground state and the corresponding energy are

$$\phi_0(x) = \left(\frac{\alpha}{\pi}\right)^{1/4} e^{-\alpha x^2/2} \, , \quad E_0 = \frac{1}{2}\hbar\omega \, ; \quad \alpha \equiv \frac{m\omega}{\hbar} \, . \quad (10.43)$$

Because the perturbing potential $V_1(x)$ is antisymmetric, the first order correction, Eq. (10.39), vanishes. To calculate the second order correction to the energy, we first find $R_1(x)$ using Eq. (10.40). We get

$$R_1(x) = \frac{2m}{\hbar^2} e^{\alpha x^2} \int_{-\infty}^{x} \epsilon \, y \, e^{-\beta y^2} e^{-\alpha y^2} dy = -\frac{m\epsilon}{\hbar^2} \frac{e^{-\beta x^2}}{\alpha + \beta} \, . \quad (10.44)$$

Then, Eq. (10.41) gives

$$\begin{aligned} E^{(2)} &= -\frac{m\epsilon^2}{2\hbar^2(\alpha+\beta)^2}\left(\frac{\alpha}{\pi}\right)^{1/2}\int_{-\infty}^{\infty} e^{-(\alpha+2\beta)\, y^2} dy \\ &= -\frac{m\epsilon^2}{2\hbar^2(\alpha+\beta)^2}\left(\frac{\alpha}{\alpha+2\beta}\right)^{1/2} . \end{aligned} \quad (10.45)$$

Note that the logarithmic perturbation method allowed for the calculation of $E^{(2)}$ in a few simple steps, without the need to calculate any series.

Exercise 10.4 Calculate Eqs. (10.44) and (10.45)

10.3 Perturbation theory: the degenerate case

In the case of degenerate system there exists multiple linearly independent eigenstates corresponding to the same eigenenergy $E_n^{(0)}$. For an N-fold degenerate eigenstate we have:

$$H_0\phi_{n,j}^{(0)} = E_n^{(0)}\phi_{n,j}^{(0)} \ ; \quad j = 1, 2, \ldots, N \ . \tag{10.46}$$

In such cases, when we calculate higher order corrections in perturbation theory, we are faced with expressions like

$$\frac{\langle\phi_k^{(0)}|\hat{V}_1|\,\phi_n^{(0)}\rangle}{E_n^{(0)} - E_k^{(0)}}$$

which diverge if there are several different states corresponding to the same energy.

To address this difficulty, we remember that when we calculated the perturbed state ψ_n, we assumed that the unperturbed state $\phi_n^{(0)}$ turns smoothly into ψ_n when we turned on the perturbation. However, in the case of a degenerate system, there are N such states for the same energy $E_n^{(0)}$. Which state exactly shall we choose as the starting point? For example, if $N = 2$, we can choose $\phi_{n,1}^{(0)}$ or $\phi_{n,2}^{(0)}$, or any arbitrary combination $\alpha\,\phi_{n,1}^{(0)} + \beta\,\phi_{n,2}^{(0)}$, where $|\alpha|^2 + |\beta|^2 = 1$. Which one is the appropriate linear combination of states that lead to the correct perturbed state of the system?

To identify the appropriate linear combination of unperturbed states we rewrite Eq. (10.4) to allow for degeneracy:

$$\psi_n = \sum_{j=1}^{N} C_j\phi_{n,j}^{(0)} + \lambda\phi_n^{(1)} + \lambda^2\phi_n^{(2)} + \cdots \ , \tag{10.47}$$

and substitute it into $\hat{H}\psi_n = E_n\psi_n$. We then get for the first order in λ

$$\hat{H}_0\,\phi_n^{(1)} + \hat{V}_1\sum_{j=1}^{N} C_j\phi_{n,j}^{(0)} = E_n^{(0)}\phi_n^{(1)} + E_n^{(1)}\sum_{k=1}^{N} C_k\phi_{n,k}^{(0)} \ . \tag{10.48}$$

Without loss of generality will consider that the degenerate states are orthonormal[3]; i.e., $\langle\phi_{n,i}^{(0)}|\phi_{n,j}^{(0)}\rangle = \delta_{ij}$. If we now take the inner product of Eq. (10.48) with $\phi_{n,i}^{(0)}$, we obtain

$$\langle\phi_{n,i}^{(0)}|\hat{H}_0|\phi_n^{(1)}\rangle + \sum_{j=1}^{N}\langle\phi_{n,i}^{(0)}|\hat{V}_1|\phi_{n,j}^{(0)}\rangle\, C_j = E_n^{(0)}\,\langle\phi_{n,i}^{(0)}|\phi_n^{(1)}\rangle + E_n^{(1)}\sum_{k=1}^{N}\langle\phi_{n,i}^{(0)}|\phi_{n,k}^{(0)}\rangle\, C_k \ .$$

Taking into account the orthonormality of $\phi_{n,i}^{(0)}$, this simplifies further to

$$\sum_{j=1}^{N}\langle\phi_{n,i}^{(0)}|\hat{V}_1|\phi_{n,j}^{(0)}\rangle\, C_j = E_n^{(1)}C_i \ , \quad i = 1, 2, \ldots, N \ . \tag{10.49}$$

[3]If the degenerate states are not orthonormal, we can always orthonormalize them via the Gram-Schmidt procedure.

Eq. (10.49) can be cast in a matrix form

$$\begin{pmatrix} \mathcal{V}_{1,1} & \mathcal{V}_{1,2} & \cdots & \mathcal{V}_{1,N} \\ \mathcal{V}_{2,1} & \mathcal{V}_{2,2} & \cdots & \mathcal{V}_{2,N} \\ \vdots & \vdots & \cdots & \vdots \\ \mathcal{V}_{N,1} & \mathcal{V}_{N,2} & \cdots & \mathcal{V}_{N,N} \end{pmatrix} \begin{pmatrix} C_1 \\ C_2 \\ \vdots \\ C_N \end{pmatrix} = E_n^{(1)} \begin{pmatrix} C_1 \\ C_2 \\ \vdots \\ C_N \end{pmatrix} , \tag{10.50}$$

where we denoted by $\mathcal{V}_{ij} = \langle \phi_{n,i}^{(0)} | \hat{V}_1 | \phi_{n,j}^{(0)} \rangle$. This is an eigenvalue problem, which when solved, gives N eigenvectors $C^k, k = 1 \ldots N$, that determine the appropriate linear combinations of $\phi_{n,j}^{(0)}$ and the corresponding energy correction $E_n^{k(1)}$ for each combination. Thus, in determining the first-order energy shifts, we are in fact diagonalizing the perturbing Hamiltonian $\hat{V}_1$ in the space spanned by the degenerate states $\phi_{n,i}^{(0)}$.

Example 10.3 Stark effect: perturbing the first excited state
As an example of a degenerate case, we return to the Stark effect in the hydrogen atom and calculate what happens to the first excited state when we apply a small electric field $\vec{\mathcal{E}}$ that perturbs the original Hamiltonian

$$\hat{H}_0 = -\frac{\hbar^2}{2m}\nabla^2 - \frac{e^2}{4\pi\epsilon_0}\frac{1}{r} ,$$

with an interaction potential

$$\hat{V}_1 = e\vec{\mathcal{E}} \cdot \hat{\vec{r}} = e\mathcal{E}\hat{z} .$$

The first excited state corresponds to the principal quantum number $n = 2$ and is four-fold degenerate. The subspace of degenerate states is spanned by $|2,0,0\rangle, |2,1,-1\rangle, |2,1,0\rangle$ and $|2,1,1\rangle$. In this basis[4], the matrix of the perturbing Hamiltonian is given by

$$e\mathcal{E}\begin{pmatrix} \langle 2,0,0|\hat{z}|2,0,0\rangle & \langle 2,0,0|\hat{z}|2,1,-1\rangle & \langle 2,0,0|\hat{z}|2,1,0\rangle & \langle 2,0,0|\hat{z}|2,1,1\rangle \\ \langle 2,1,-1|\hat{z}|2,0,0\rangle & \langle 2,1,-1|\hat{z}|2,1,-1\rangle & \langle 2,1,-1|\hat{z}|2,1,0\rangle & \langle 2,1,-1|\hat{z}|2,1,1\rangle \\ \langle 2,1,0|\hat{z}|2,0,0\rangle & \langle 2,1,0|\hat{z}|2,1,-1\rangle & \langle 2,1,0|\hat{z}|2,1,0\rangle & \langle 2,1,0|\hat{z}|2,1,1\rangle \\ \langle 2,1,1|\hat{z}|2,0,0\rangle & \langle 2,1,1|\hat{z}|2,1,-1\rangle & \langle 2,1,1|\hat{z}|2,1,0\rangle & \langle 2,1,1|\hat{z}|2,1,1\rangle \end{pmatrix}$$

The computation of the matrix elements $\mathcal{V}_{ij}$ can be significantly simplified by using symmetry considerations. Because the electric field points in the z direction, $\hat{V}_1$ is invariant under rotations about the z axis, and therefore $\hat{V}_1$ commutes with $\hat{L}_z$ —the z component of the angular momentum operator. This implies on one hand, that

$$\langle n,l',m'|\hat{L}_z\hat{z}|n,l,m\rangle = m' \langle n,l',m'|\hat{z}|n,l,m\rangle ,$$

and, on the other hand

$$\langle n,l',m'|\hat{L}_z\hat{z}|n,l,m\rangle = \langle n,l',m'|\hat{z}\hat{L}_z|n,l,m\rangle = m \langle n,l',m'|\hat{z}|n,l,m\rangle .$$

[4]The order of vectors in the basis is irrelevant, as it won't change the eigenvalue problem.

Thus, if $m \neq m'$

$$\langle n, l', m' | \hat{z} | n, l, m \rangle = 0 \; . \tag{10.51}$$

Consequently, all matrix elements with $m \neq m'$ vanish. Additionally, all diagonal elements must vanish as well, because they are integrals of odd functions over the entire space.

We are left with a much simplified perturbation matrix

$$e\mathcal{E} \begin{pmatrix} 0 & 0 & \langle 2,0,0|\hat{z}|2,1,0\rangle & 0 \\ 0 & 0 & 0 & 0 \\ \langle 2,1,0|\hat{z}|2,0,0\rangle & 0 & 0 & 0 \\ 0 & 0 & 0 & 0 \end{pmatrix} ,$$

in which we can calculate the remaining non-zero elements. Remember that $\langle r, \theta, \varphi | n, l, m \rangle = R_{nl}(r) Y_{lm}(\theta, \varphi)$. Using the radial wave functions R_{20} and R_{21} together with the spherical harmonics Y_{00} and Y_{10}, and writing z as $r\cos\theta$, we have

$$\begin{aligned} \langle 2,0,0|\hat{z}|2,1,0\rangle &= \int_0^\infty r^2 dr \int_0^\pi \sin(\theta) d\theta \int_0^{2\pi} d\varphi \, \overline{R}_{20} \overline{Y}_{00} \, r\cos(\theta) R_{21} Y_{10} \\ &= -3a_0 \qquad\qquad\qquad\qquad\qquad\qquad\qquad\qquad (10.52) \\ &= \langle 2,1,0|\hat{z}|2,0,0\rangle \; , \end{aligned}$$

where a_0 is Bohr's radius. Therefore, the first-order perturbation matrix for the Stark effect in hydrogen, for the first excited states, becomes

$$\begin{pmatrix} 0 & 0 & -3a_0 e\,\mathcal{E} & 0 \\ 0 & 0 & 0 & 0 \\ -3a_0 e\,\mathcal{E} & 0 & 0 & 0 \\ 0 & 0 & 0 & 0 \end{pmatrix} .$$

Next, we find the appropriate linear combinations of the degenerate states, and the corresponding first-order energy shifts. For this, we have to solve the eigenvalue problem

$$\begin{pmatrix} 0 & 0 & -3a_0 e\,\mathcal{E} & 0 \\ 0 & 0 & 0 & 0 \\ -3a_0 e\,\mathcal{E} & 0 & 0 & 0 \\ 0 & 0 & 0 & 0 \end{pmatrix} \begin{pmatrix} C_1 \\ C_2 \\ C_3 \\ C_4 \end{pmatrix} = E_2^{(1)} \begin{pmatrix} C_1 \\ C_2 \\ C_3 \\ C_4 \end{pmatrix} . \tag{10.53}$$

Eq. (10.53) admits non-trivial solutions only if

$$\begin{vmatrix} -E_2^{(1)} & 0 & -3a_0 e\,\mathcal{E} & 0 \\ 0 & -E_2^{(1)} & 0 & 0 \\ -3a_0 e\,\mathcal{E} & 0 & -E_2^{(1)} & 0 \\ 0 & 0 & 0 & -E_2^{(1)} \end{vmatrix} = 0 \; ,$$

which yields the first-order energy shifts in the energy

$$E_2^{(1)} = 0 \; , \quad 0 \; , \quad 3a_0 e\,\mathcal{E} \; , \quad -3a_0 e\,\mathcal{E} \; . \tag{10.54}$$

Substituting these back into Eq. (10.53) we get the corresponding eigenvectors, which are the sought after linear combinations of degenerate eigenstates:

$$|2,1,1\rangle\,,\quad |2,1,-1\rangle\,,\quad \frac{1}{\sqrt{2}}\big(\,|2,0,0\rangle - |2,1,0\rangle\big),\ \frac{1}{\sqrt{2}}\big(\,|2,0,0\rangle + |2,1,0\rangle\big)\,.$$

Their corresponding energies are illustrated in Fig. 10.1. Note that due to

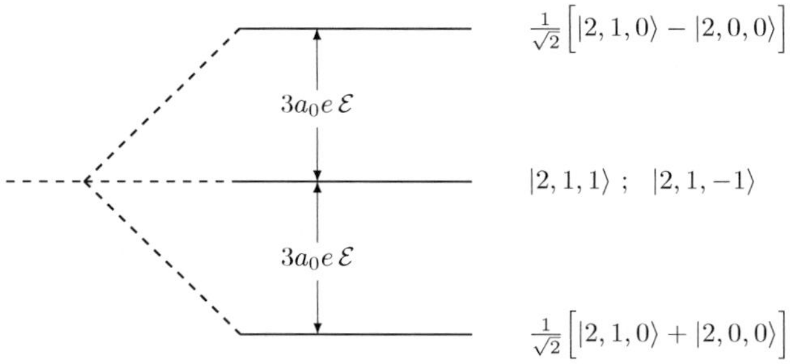

FIGURE 10.1
Stark Effect —the degenerate case: first-order shifts in the hydrogen energy levels for the first excited state.

the axial z symmetry, which leads to $[\hat{V}_1, \hat{L}_z] = 0$, only the states with the same values of m are mixed together by the perturbation. The symmetric and anti-symmetric linear combinations of $|2,0,0\rangle$ and $|2,1,0\rangle$ no longer have definite parity, even if each individual state has. Therefore, they have a non-zero expectation value of the electric dipole moment, which allows them to interact with the external electric field, thus partially lifting the degeneracy. In this case, we do have a linear Stark effect; i.e., the energy shift is linear in the applied external electric field.

Exercise 10.5 Find the eigenvectors and eigenenergies for Eq. (10.53).

10.4 Problems

10.1 A charged particle in a one-dimensional harmonic oscillator potential is subject to a week electric field $\mathcal{E}$ which shifts the potential energy by $V_1(x) = -q\mathcal{E}x$. Calculate the first and second order corrections in the energy levels of the particle.

10.2 Solve Problem 10.1 using the logarithmic perturbation method. Calculate the second order correction only for the ground state energy.

10.3 Consider an electron in an infinite one-dimensional potential well subject to a uniform electric field of magnitude $\mathcal{E}$ as illustrated in Fig. 10.2. Find

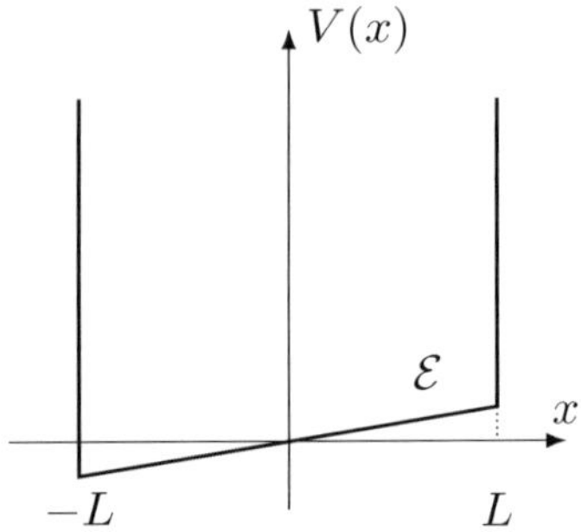

FIGURE 10.2
Figure for Problem 10.3: perturbing the infinite square well.

the first non-zero correction to the ground state energy.

10.4 A particle of mass m is subject to a potential

$$V(x) = \frac{1}{2}m\omega^2 x^2 \left(\frac{x}{a}\right)^{2\beta}$$

where a is a constant with dimensions of length, and $\beta \ll 1$ is a dimensionless exponent. This potential is an example of an anharmonic oscillator, which is a potential very close to the harmonic oscillator potential. Therefore, it can be rewritten it as

$$V(x) = \frac{1}{2}m\omega^2 x^2 + V_1(x) ,$$

where

$$V_1(x) = \frac{1}{2}m\omega^2 x^2 \left[\left(\frac{x}{a}\right)^{2\beta} - 1\right] .$$

Considering V_1 as a small perturbation, approximate the ground state of the anharmonic oscillator by calculating the first-order correction to the ground state energy of the harmonic oscillator.

10.5 Calculate the matrix element $\langle 2,1,0|\hat{z}|2,0,0\rangle$ of Eq. (10.52) using the explicit expressions for the radial functions and the spherical harmonics, respectively:

$$R_{20}(r) = 2\left(\frac{1}{2a_0}\right)^{3/2}\left(1 - \frac{r}{2a_0}\right)e^{-r/2a_0} ,$$

$$R_{21}(r) = \frac{1}{\sqrt{3}} \left(\frac{1}{2a_0} \right)^{3/2} \frac{r}{a_0} e^{-r/2a_0} \,,$$

$$Y_{10}(\theta, \varphi) = \sqrt{\frac{3}{4\pi}} \cos\theta \,, \quad Y_{00}(\theta, \varphi) = \sqrt{\frac{1}{4\pi}} \,.$$

10.6 A particle of mass m in a two-dimensional oscillator potential

$$V(x, y) = \frac{1}{2} m\omega^2 (x^2 + y^2)$$

is subject to a perturbation

$$V_1(x, y) = \varepsilon\, m\omega^2 xy \,,$$

where $\varepsilon \ll 1$ is a dimensionless parameter. The unperturbed first-excited state has the energy $E_1^{(0)} = 2\hbar\omega$ and is two-fold degenerate. The corresponding subspace of degenerate states is spanned by

$$\phi_1(x, y) = \phi_0(x)\phi_1(y) \,, \quad \phi_2(x, y) = \phi_1(x)\phi_0(y) \,,$$

where ϕ_0 and ϕ_1 are the one-dimensional harmonic oscillator states. Calculate the first-order energy shifts and the proper linear combinations of degenerate eigenstates.

10.7 A particle of mass m is free to move on a circular path of radius R.

(a) Show that all states —with the exception of the ground state, are doubly degenerate.

(b) If we introduce a perturbation $V_1(s) = -Ae^{-s^2/a^2}$ where $a \ll R$ and s is the coordinate along the circumference, calculate the first-order correction to energy $E_1^{(0)}$.

11

Time Dependent Perturbation Theory

11.1 Introduction

In time independent perturbation theory we were interested in finding the eigenvalues and eigenstates of a system whose Hamiltonian could be written as the sum of a known Hamiltonian and another term, describing the "perturbation." The idea was that if the perturbation was "small" enough, then the departure of the eigenvalues and eigenstates of the perturbed Hamiltonian from the unperturbed case, would be small too. The formalism introduced a small arbitrary parameter λ, which multiplied the perturbative term in the Hamiltonian. Then, with the hope of convergence, in the Schrödinger equation we expanded the eigenenergies and eigenstates in powers of λ, and required that the coefficients of the same powers of λ on both sides of this equation be equal. The set of equations thus obtained, allowed for iterative solutions in each order of λ. The small parameter λ was used both to justify the series expansion and to keep track of various orders in the expansion. At the end of the calculation, λ was set equal to 1 and we thus recovered the original Hamiltonian.

In time dependent perturbation theory the goal is different. We want to find the time evolution operator of the perturbed system, and use it to estimate the transition probabilities between two states of the unperturbed system. We could also use it to estimate the decay probability by modeling the irreversible process via the coupling of a small system, the original system, to a very large quantum system.

In the case of a time independent Hamiltonian $\hat{H}_0$, the evolution of an arbitrary quantum state is described by the time evolution operator

$$\hat{U}(t,0) = e^{-i\hat{H}_0 t/\hbar}$$

and thus

$$|\Psi(t)\rangle = e^{-i\hat{H}_0 t/\hbar}|\Psi(0)\rangle = \sum_n e^{-iE_n t/\hbar} c_n |n\rangle \ ,$$

where $|n\rangle$ are the eigenstates Hamiltonian, $\hat{H}_0|n\rangle = E_n|n\rangle$.

However, such a formalism doesn't describe interactions that depend on time

$$i\hbar\frac{d}{dt}|\psi(t)\rangle = \hat{H}(t)|\psi(t)\rangle \ . \tag{11.1}$$

DOI: 10.1201/9781003174967-11

To apply the time dependent perturbation theory we assume that the Hamiltonian of the system can be expressed as

$$\hat{H}(t) = \hat{H}_0 + \hat{\mathcal{V}}(t) , \tag{11.2}$$

where $\hat{H}_0$ is a known time independent Hamiltonian and $\hat{\mathcal{V}}(t)$ is a small perturbation. Similarly to the time independent case, we could keep track of the perturbation order using a small parameter λ:

$$\hat{H}(t) = \hat{H}_0 + \lambda \hat{\mathcal{V}}(t) . \tag{11.3}$$

At the end of the calculation we would then set $\lambda = 1$ and recover the initial Hamiltonian.

Of course, we can substitute directly Eq. (11.3) into Eq. (11.1) and try to solve order by order the resulting equations. However, such an approach could lead to equations which are difficult to solve. It turns out that in this cases it is more effective to use another representation of quantum mechanics, the *interaction picture.*

11.2 The interaction picture

The interaction picture is an intermediate representation of quantum mechanics, which can be seen as a picture in between the Schrödinger and Heisenberg pictures. In the interaction picture, both the state vectors and the operators change with time. Therefore, we will find the time evolution equations for each.

Consider the Hamiltonian $\hat{H}(t) = \hat{H}_0 + \hat{\mathcal{V}}(t)$ where the time dependence resides in the potential $\hat{\mathcal{V}}(t)$. In the Schrödinger picture, the dynamics of the system is described by the Schrödinger equation $i\hbar\frac{d}{dt}|\Psi_S(t)\rangle = \hat{H}|\Psi_S(t)\rangle$. We define the quantum state in the interaction picture by

$$|\Psi_I(t)\rangle = e^{i\hat{H}_0 t/\hbar}|\Psi_S(t)\rangle , \tag{11.4}$$

where $|\Psi_I(0)\rangle = |\Psi_S(0)\rangle$. To find the equation of motion for the state $|\Psi_I(t)\rangle$ in the interaction picture, we take the time derivative of Eq. (11.4). We obtain

$$\begin{aligned} i\hbar\frac{d}{dt}|\Psi_I(t)\rangle &= -\hat{H}_0 e^{i\hat{H}_0 t/\hbar}|\Psi_S(t)\rangle + e^{i\hat{H}_0 t/\hbar} i\hbar\frac{d}{dt}|\Psi_S(t)\rangle \\ &= -\hat{H}_0 e^{i\hat{H}_0 t/\hbar}|\Psi_S(t)\rangle + e^{i\hat{H}_0 t/\hbar}\left(\hat{H}_0 + \hat{\mathcal{V}}(t)\right)|\Psi_S(t)\rangle \\ &= e^{i\hat{H}_0 t/\hbar}\hat{\mathcal{V}}(t)|\Psi_S(t)\rangle = e^{i\hat{H}_0 t/\hbar}\hat{\mathcal{V}}(t)\, e^{-i\hat{H}_0 t/\hbar}|\Psi_I(t)\rangle \\ &= \hat{\mathcal{V}}_I(t)|\Psi_I(t)\rangle , \end{aligned}$$

where $\hat{\mathcal{V}}_I(t) = e^{i\hat{H}_0 t/\hbar}\hat{\mathcal{V}}(t)\, e^{-i\hat{H}_0 t/\hbar}$.

Thus, in the interaction picture, the quantum state of the system obeys a Schrödinger-like equation of motion where only the small, external, time-dependent term $\hat{\mathcal{V}}_I(t)$ is present:

$$i\hbar\frac{d}{dt}\,|\Psi_I(t)\rangle = \hat{\mathcal{V}}_I(t)\,|\Psi_I(t)\rangle \ . \tag{11.5}$$

Having established the equation of motion for the quantum states, we now turn our attention to operators. The expectation value of an operator can be written as

$$\langle\Psi_S(t)|\hat{A}_S|\Psi_S(t)\rangle = \langle\Psi_I(t)|e^{i\hat{H}_0t/\hbar}\hat{A}_S\,e^{-i\hat{H}_0t/\hbar}|\Psi_I(t)| = \langle\Psi_I(t)|\hat{A}_I|\Psi_I(t)\rangle \ ,$$

where we defined the operator in the interaction picture as

$$\hat{A}_I = e^{i\hat{H}_0t/\hbar}\hat{A}_S\,e^{-i\hat{H}_0t/\hbar} \ . \tag{11.6}$$

If $\hat{A}_S$ doesn't depend on time, then the time dependence of $\hat{A}_I$ is given by

$$\begin{aligned}\frac{d\hat{A}_I}{dt} &= \frac{i}{\hbar}\left(e^{i\hat{H}_0t/\hbar}\hat{H}_0\hat{A}_S\,e^{-i\hat{H}_0t/\hbar} - e^{i\hat{H}_0t/\hbar}\hat{A}_S\hat{H}_0\,e^{-i\hat{H}_0t/\hbar}\right)\\ &= \frac{i}{\hbar}\left(\hat{H}_0\,e^{i\hat{H}_0t/\hbar}\hat{A}_S\,e^{-i\hat{H}_0t/\hbar} - e^{i\hat{H}_0t/\hbar}\hat{A}_S\,e^{-i\hat{H}_0t/\hbar}\hat{H}_0\right)\\ &= \frac{i}{\hbar}\left[\hat{H}_0, \hat{A}_I\right] \ .\end{aligned} \tag{11.7}$$

Thus, in the interaction picture, both the quantum states and the operators change with time. Collecting the above results we have:

$$i\hbar\frac{d}{dt}\,|\Psi_I(t)\rangle = \hat{\mathcal{V}}_I(t)\,|\Psi_I(t)\rangle \ , \tag{11.8}$$

$$\frac{d\hat{A}_I}{dt} = \frac{i}{\hbar}\left[\hat{H}_0, \hat{A}_I\right] \ . \tag{11.9}$$

Note that while the time evolution of quantum states in the interaction picture is determined by $\hat{\mathcal{V}}_I(t)$, the evolution of operators is governed by $\hat{H}_0$.

11.2.1 The time evolution operator in the interaction picture

It is often more advantageous to calculate the time evolution operator $\hat{U}$ instead of the time evolution of the state vector $|\Psi(t)\rangle$, because once we know $\hat{U}$ we can find the quantum state at time t for any initial state.

In the interaction picture, the state $|\Psi_I(t)\rangle$ can be related to the state at $t = t_0$ via the time evolution operator $\hat{U}_I(t, t_0)$:

$$|\Psi_I(t)\rangle = \hat{U}_I(t, t_0)\,|\Psi_I(t_0)\rangle \ . \tag{11.10}$$

Without loss of generality we can choose $t_0 = 0$, and henceforth we will write simply $\hat{U}_I(t)$ instead of $\hat{U}_I(t, 0)$.

Because Eq. (11.10) is true for any initial state $|\Psi_I(0)\rangle$, we must have

$$i\hbar\frac{d}{dt}\hat{U}_I(t) = \hat{\mathcal{V}}_I(t)\hat{U}_I(t) \ , \tag{11.11}$$

with the initial condition $\hat{U}_I(0) = 1$.

We can obtain a formal solution for $\hat{U}_I(t)$ by integrating Eq. (11.11) repeatedly. We first integrate it from 0 to t and obtain the equation

$$\hat{U}_I(t) = 1 - \frac{i}{\hbar}\int_0^t d\tau_1 \hat{\mathcal{V}}_I(\tau_1)\hat{U}_I(\tau_1) \ .$$

If we use this expression for $\hat{U}_I(\tau_1)$ under the integral sign above, we obtain

$$\begin{aligned}\hat{U}_I(t) &= 1 + \left(-\frac{i}{\hbar}\right)\int_0^t d\tau_1 \hat{\mathcal{V}}_I(\tau_1) \\ &+ \left(-\frac{i}{\hbar}\right)^2 \int_0^t d\tau_1 \int_0^{\tau_1} d\tau_2\, \hat{\mathcal{V}}_I(\tau_1)\hat{\mathcal{V}}_I(\tau_2)\hat{U}_I(\tau_2) \ ,\end{aligned}$$

where $0 \leq \tau_2 \leq \tau_1 \leq t$. This could be further iterated to get

$$\begin{aligned}\hat{U}_I(t) &= 1 + \left(-\frac{i}{\hbar}\right)\int_0^t d\tau_1 \hat{\mathcal{V}}_I(\tau_1) \\ &+ \left(-\frac{i}{\hbar}\right)^2 \int_0^t d\tau_1 \int_0^{\tau_1} d\tau_2\, \hat{\mathcal{V}}_I(\tau_1)\hat{\mathcal{V}}_I(\tau_2) \\ &+ \left(-\frac{1}{\hbar}\right)^3 \int_0^t d\tau_1 \int_0^{\tau_1} d\tau_2 \int_0^{\tau_2} d\tau_3\, \hat{\mathcal{V}}_I(\tau_1)\hat{\mathcal{V}}_I(\tau_2)\hat{\mathcal{V}}_I(\tau_3)\hat{U}_I(\tau_3) \ ,\end{aligned}$$

where $0 \leq \tau_3 \leq \tau_2 \leq \tau_1 \leq t$. If we continue the iteration process we arrive at a formal series solution:

$$\hat{U}_I(t) = \sum_{n=0}^{\infty}\left(-\frac{i}{\hbar}\right)^n \int_0^t d\tau_1 \cdots \int_0^{\tau_{n-1}} d\tau_n \hat{\mathcal{V}}_I(\tau_1)\cdots\hat{\mathcal{V}}_I(\tau_{n-1})\hat{\mathcal{V}}_I(\tau_n) \ . \tag{11.12}$$

Here, the term $n = 0$ gives 1, and $0 \leq \tau_n \leq \tau_{n-1} \leq \cdots \leq \tau_1 \leq t$. If the perturbation is small, this series expansion could be considered as adding smaller and smaller terms. The hope is that, if the series converges sufficiently fast, retaining only the first few terms gives a good enough approximation for the time evolution operator. We note that the operators $\hat{\mathcal{V}}_I$ in Eq. (11.12) are arranged in a time-ordered sequence, with $0 \leq \tau_n \leq \tau_{n-1} \leq \cdots \leq \tau_1 \leq t$, starting with the most recent operator.

The problem is now "solved." We still need to interpret the result, and attempt to compute the *transition probabilities* by putting together these pieces. To do so, we first switch back to the Schrödinger picture. Using $|\psi_I(0)\rangle = |\psi_S(0)\rangle$ together with Eqs. (11.4) and (11.10), we have

$$\begin{aligned}|\psi_I(t)\rangle &= e^{i\hat{H}_0 t/\hbar}|\psi_S(t)\rangle = e^{i\hat{H}_0 t/\hbar}\hat{U}_S(t)|\psi_S(0)\rangle \\ &= \hat{U}_I(t)|\psi_I(0)\rangle = \hat{U}_I(t)|\psi_S(0)\rangle \ .\end{aligned}$$

Thus, the time evolution operator in Schrödinger picture becomes

$$\hat{U}_S(t) = e^{-i\hat{H}_0 t/\hbar}\,\hat{U}_I(t) \equiv \hat{U}_0(t)\hat{U}_I(t)\ , \tag{11.13}$$

where $\hat{U}_0(t) \equiv e^{-i\hat{H}_0 t/\hbar}$. Note that $\hat{U}_0(t_1)\hat{U}_0^\dagger(t_2) = \hat{U}_0(t_1 - t_2)$. Then, taking into account that $\hat{\mathcal{V}}_I(t) = \hat{U}_0^\dagger(t)\,\hat{\mathcal{V}}(t)\,\hat{U}_0(t)$, Eq. (11.12) can be written as

$$\begin{aligned}
\hat{U}_S(t) &= \hat{U}_0(t) + \left(-\frac{i}{\hbar}\right)\int_0^t d\tau_1 \hat{U}_0(t-\tau_1)\hat{\mathcal{V}}(\tau_1)\hat{U}_0(\tau_1) \\
&+ \left(-\frac{i}{\hbar}\right)^2 \int_0^t d\tau_1 \int_0^{\tau_1} d\tau_2\, \hat{U}_0(t-\tau_1)\hat{\mathcal{V}}(\tau_1)\hat{U}_0(\tau_1-\tau_2)\hat{\mathcal{V}}(\tau_2)\hat{U}_0(\tau_2) \\
&+ \left(-\frac{1}{\hbar}\right)^3 \int_0^t d\tau_1 \int_0^{\tau_1} d\tau_2 \int_0^{\tau_2} d\tau_3\, \hat{U}_0(t-\tau_1)\hat{\mathcal{V}}(\tau_1)\hat{U}_0(\tau_1-\tau_2)\hat{\mathcal{V}}(\tau_2)\hat{U}_0(\tau_2-\tau_3)\hat{U}_0(\tau_3) \\
&+ \cdots
\end{aligned} \tag{11.14}$$

where, as before, $0 \le \tau_n \le \tau_{n-1} \le \cdots \le \tau_1 \le t$.

To interpret this result, imagine that we apply the above time evolution operator on an initial state $|\psi_S(0)\rangle$. The first term of Eq. (11.14) describes

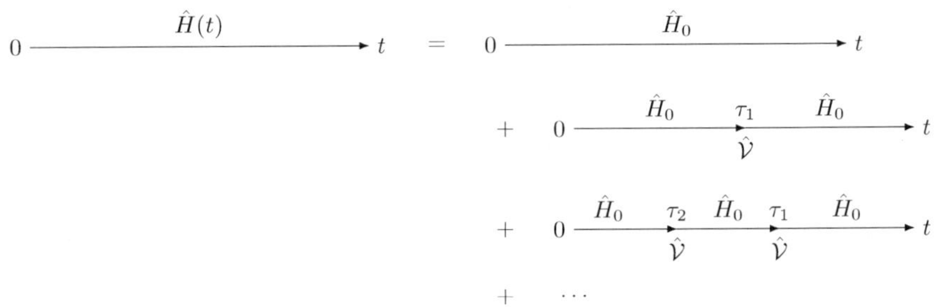

FIGURE 11.1
A pictorial representation of the time-dependent perturbative expansion.

simply the evolution of the unperturbed system, under the action of the time independent Hamiltonian $\hat{H}_0$. We will call this "free evolution." This process is illustrated in Fig. 11.1.

The second term describes the free evolution for a time τ_1, followed by a sudden jump from $|\psi_S(\tau_1)\rangle$ to a new state $\hat{\mathcal{V}}(\tau_1)\,|\psi_S(\tau_1)\rangle$. Then, the system evolves freely from τ_1 to t. The integration over τ_1 is in fact a sum over all possible "paths" from 0 to t that contain exactly one transition. Similarly, the next term is a sum of all paths with two transitions, the third one a sum of all paths with three transitions, and so on. In this way, we see that the time evolution operator for a time dependent Hamiltonian system can be interpreted as a sum over all possible quantum transition sequences from the initial time 0 to t.

11.3 Transition probabilities

As we have already seen in Ch. 9, if a particle is in state $|\psi\rangle$, the probability to find it in state $|\phi\rangle$ is

$$P = |\langle\phi|\psi\rangle|^2 \ .$$

Consider now two eigenstates of $\hat{H}_0$ denoted by $|i\rangle$ and $|f\rangle$, with eigenenergies $E_i = \hbar\omega_i$ and $E_f = \hbar\omega_f$. Assume that at $t = 0$ the particle is in state $|i\rangle$. We are interested in calculating the transition probability $P_{f\leftarrow i}(t)$ that at a later time t we would find the particle in state $|f\rangle$. For this we calculate the matrix element $\langle f|\hat{U}_S(t)|i\rangle$ of the evolved state $\hat{U}_S(t)\,|i\rangle$ with state $|f\rangle$. This gives

$$P_{f\leftarrow i}(t) = |\langle f|\hat{U}_S(t)|i\rangle|^2 = |\langle f|\hat{U}_0(t)\hat{U}_I(t)|i\rangle|^2 \ . \tag{11.15}$$

The probability amplitude for this process is thus

$$u_{f\leftarrow i}(t) = \langle f|\hat{U}_0(t)\hat{U}_I(t)|i\rangle = e^{-i\omega_f t}\,\langle f|\hat{U}_I(t)|i\rangle \ . \tag{11.16}$$

Remember that in Eq. (11.12) we have already calculated the time evolution operator $\hat{U}_I(t)$. Thus, the probability amplitude becomes

$$\begin{aligned}
u_{f\leftarrow i}(t) &= e^{-i\omega_f t}\Big[\delta_{fi} + \left(\frac{-i}{\hbar}\right)\int_0^t d\tau_1\,\langle f|\hat{\mathcal{V}}_I(\tau_1)|i\rangle \\
&+ \left(\frac{-i}{\hbar}\right)^2 \sum_{m_2}\int_0^t d\tau_1 \int_0^{\tau_1} d\tau_2\,\langle f|\hat{\mathcal{V}}_I(\tau_1)|m_2\rangle\,\langle m_2|\hat{\mathcal{V}}_I(\tau_2)|i\rangle + \cdots \\
&+ \left(\frac{-i}{\hbar}\right)^n \sum_{m_2,\ldots,m_n}\int_0^t d\tau_1 \cdots \int_0^{\tau_{n-1}} d\tau_n\,\langle f|\hat{\mathcal{V}}_I(\tau_1)|m_2\rangle\,\langle m_2|\hat{\mathcal{V}}_I(\tau_2)|m_3\rangle \cdots \\
&\quad \langle m_{n-1}|\hat{\mathcal{V}}_I(\tau_{n-1})|m_n\rangle\,\langle m_n|\hat{\mathcal{V}}_I(\tau_n)|i\rangle + \cdots\Big] \ ,
\end{aligned} \tag{11.17}$$

where we have employed the completeness formula $\sum_n |n\rangle\,\langle n| = 1$.

Exercise 11.1 Derive Eq. (11.17).

As before, we now switch to the Schrödinger picture and write

$$\langle f|\hat{\mathcal{V}}_I(\tau_1)|i\rangle = \langle f|\hat{U}_0^\dagger(\tau_1)\hat{\mathcal{V}}_S(\tau_1)\hat{U}_0(\tau_1)|i\rangle = e^{i\omega_{fi}\tau_1}\mathcal{V}_{lk}(\tau_1) \ , \tag{11.18}$$

where, for a slightly more convenient notation, we defined

$$\mathcal{V}_{fi}(\tau_1) \equiv \langle f|\hat{\mathcal{V}}_S(\tau_1)|i\rangle \quad ; \quad \omega_{fi} \equiv \omega_f - \omega_i \ .$$

With this, the n-th order term in the perturbative expansion of Eq. (11.17) can be written as

$$u^{(n)}_{f\leftarrow i}(t) = e^{-i\omega_f t}\left(\frac{-i}{\hbar}\right)^n \sum_{m_2,\ldots,m_n} \int_0^t d\tau_1 \int_0^t d\tau_2 \cdots \int_0^{\tau_{n-1}} d\tau_n \mathcal{V}_{fm_2}(\tau_1)\, e^{i\omega_{fm_2}\tau_1}$$
$$\times\, \mathcal{V}_{m_2m_3}(\tau_2)\, e^{i\omega_{m_2m_3}\tau_2} \cdots \mathcal{V}_{m_ni}(\tau_n)\, e^{i\omega_{m_ni}\tau_n}\ . \tag{11.19}$$

Note that the sums are over all possible intermediate states, whereas the integrals are over all possible intermediate transition times.

One last observation: in practical cases is enough to calculate only the lower order terms. For example, the first three terms are:

$$\text{order } 0: \quad u^{(0)}_{f\leftarrow i}(t) = e^{-i\omega_f t}\delta_{fi}$$
$$\text{order } 1: \quad u^{(1)}_{f\leftarrow i}(t) = e^{-i\omega_f t}\left(\frac{-i}{\hbar}\right)\int_0^t d\tau_1 \mathcal{V}_{fi}(\tau_1)e^{i\omega_{fi}\tau_1}$$
$$\text{order } 2: \quad u^{(2)}_{f\leftarrow i}(t) = e^{-i\omega_f t}\left(\frac{-i}{\hbar}\right)^2 \sum_{m_2}\int_0^t d\tau_1 \int_0^{\tau_1} d\tau_2$$
$$\mathcal{V}_{fm_2}(\tau_1)e^{i\omega_{fm_2}\tau_1}\mathcal{V}_{m_2i}(\tau_2)e^{i\omega_{m_2i}\tau_2}$$

We arrived at our final step, calculating the transition probability. The transition probability that at time t the system is found in state $|f\rangle$ if at $t = 0$ it was in state $|i\rangle$, is given by:

$$P_{f\leftarrow i}(t) = |\delta_{fi} + u^{(1)}_{f\leftarrow i}(t) + u^{(2)}_{f\leftarrow i}(t) + \cdots|^2\ . \tag{11.20}$$

Note that the transition probability is not the sum of the squares of the absolute values of the amplitudes. It is the square of the absolute value of their sum. This is an important distinction, because different pathways from $|i\rangle$ to $|f\rangle$ can interfere with each other either constructively or destructively. This quantum interference will affect the transition probability.

To calculate the transition probability we square the amplitude. Therefore, if we retain only the first order in the expansion of the transition amplitude, we already obtain the transition probability to the second order. This is a good approximation for many practical cases. For $i \neq f$, we have

$$P_{f\leftarrow i}(t) = |u^{(1)}_{f\leftarrow i}(t)|^2 + (\text{higher orders...}) \tag{11.21}$$
$$= \frac{1}{\hbar^2}\left|\int_0^t d\tau_1 \mathcal{V}_{fi}(\tau_1)e^{i\omega_{fi}\tau_1}\right|^2 + (\text{higher orders...})\ .$$

For $i = f$, we use the conservation of probability

$$P_{i\leftarrow i}(t) = 1 - \sum_{f\neq i}|u^{(1)}_{f\leftarrow i}(t)|^2 + (\text{higher orders...})\ . \tag{11.22}$$

Henceforth, we are going to ignore the higher orders, and simply write

$$P_{f\leftarrow i}(t) = \frac{1}{\hbar^2}\left|\int_0^t d\tau_1 \mathcal{V}_{fi}(\tau_1)e^{i\omega_{fi}\tau_1}\right|^2 , \tag{11.23}$$

and

$$P_{i\leftarrow i}(t) = 1 - \sum_{f\neq i} |u^{(1)}_{f\leftarrow i}(t)|^2 \tag{11.24}$$

respectively.

Example 11.1 Perturbing the harmonic oscillator

A particle of mass m, initially in the ground state of a harmonic oscillator potential $V(x) = m\omega^2x^2/2$, is acted upon by a perturbation $\hat{\mathcal{V}}(t) = D\,\hat{x}\,e^{-\beta t}$ for a duration $0 \leq t < \infty$. Here D and β constants with appropriate dimensions. The perturbative potential can be seen as a time-dependent electric potential acting on a charged particle. Calculate the probability of finding the particle in the first excited state. Show that for large values of t we obtain a constant transition probability.

To answer this question, we will first consider the unperturbed problem. A particle of mass m in the harmonic potential $V(x) = m\omega^2x^2/2$ has

$$E_n = \left(n + \frac{1}{2}\right)\hbar\omega \; ; \;\; \psi_n(x) = \left(\frac{m\omega}{\pi\hbar}\right)^{1/4}\frac{1}{\sqrt{2^n n!}}H_n(\xi)e^{-\xi^2/2} , \quad n = 0, 1, 2, \ldots$$

where $\xi = x\sqrt{m\omega/\hbar}$ and H_n is the Hermite polynomial of order n. The ground state corresponds to $n = 0$ and the first excited state to $n = 1$. Using Eq. (11.23) we can write

$$\begin{aligned} P_{1\leftarrow 0}(t) &= \frac{1}{\hbar^2}\left|\int_0^t d\tau \mathcal{V}_{10}(\tau)e^{i\omega_{10}\tau}\right|^2 = \frac{1}{\hbar^2}\left|\int_0^t d\tau\, \langle 1|\hat{\mathcal{V}}(\tau)|0\rangle\, e^{i\omega\tau}\right|^2 \\ &= \frac{D^2}{\hbar^2}\,|\,\langle 1|\hat{x}|0\rangle\,|^2\left|\int_0^t e^{(i\omega-\beta)\tau}d\tau\right|^2 , \end{aligned}$$

where, for the harmonic oscillator, $\omega_{10} = (E_1 - E_0)/\hbar = \omega$. Then we calculate

$$\langle 1|\hat{x}|0\rangle = \int_{-\infty}^{\infty} x\,\psi_1^*(x)\psi_0(x)\,dx = \frac{m\omega}{\hbar}\sqrt{\frac{2}{\pi}}\int_{-\infty}^{\infty} x^2 e^{-\frac{mx^2\omega}{\hbar}}dx = \sqrt{\frac{\hbar}{2m\omega}} ,$$

and

$$\left|\int_0^t e^{(i\omega-\beta)\tau}d\tau\right|^2 = \left|\frac{e^{i\omega-\beta}-1}{i\omega-\beta}\right|^2 = \frac{e^{-2\beta t} + 1 - 2e^{-\beta t}\cos(\omega t)}{\omega^2+\beta^2} .$$

Putting together these results, we obtain the transition probability from the ground state to the first excited state under the action of the time-dependent perturbation:

$$P_{1\leftarrow 0}(t) = \frac{D^2}{2m\hbar\omega}\left[\frac{e^{-2\beta t} + 1 - 2e^{-\beta t}\cos(\omega t)}{\omega^2+\beta^2}\right] . \tag{11.25}$$

Finally, in the limit $t \to \infty$, Eq. (11.25) reduces to

$$P_{1\leftarrow 0}(t) = \frac{D^2}{2m\hbar\omega}\left[\omega^2 + \beta^2\right]^{-1} . \tag{11.26}$$

11.4 Fermi's golden rule

How do we calculate the transition probability for the case where the initial state is in the discrete part of the energy spectrum, a bound state, while the final state is in the continuum? Consider for example the ionization of the hydrogen. The initial state is a bound state, while the final state is a free electron.

The free electron's energy is in the continuum and is described by a momentum eigenstate, which is non-normalizable. In this case we will integrate over the energy of the final state, and obtain a transition probability which is linear in time. The transition rate associated to this probability is known as *Fermi's Golden Rule.*

We will consider the case of a time dependent perturbation, where at $t = 0$ we "turn on" a constant perturbation $\hat{V}$:

$$\hat{\mathcal{V}}(t) = \begin{cases} 0 & t < 0 \\ \hat{V} & t \geq 0 \end{cases}$$

Then, Eq. (11.23) gives

$$P_{f\leftarrow i}(t) = \frac{1}{\hbar^2}\left|\langle f|\hat{V}|i\rangle \int_0^t d\tau_1 e^{i\omega_{fi}\tau_1}\right|^2 = \frac{1}{\hbar^2}\left|\langle f|\hat{V}|i\rangle \frac{e^{i\omega_{fi}t} - 1}{\omega_{fi}}\right|^2 , \tag{11.27}$$

or

$$P_{f\leftarrow i}(t) = \frac{1}{\hbar^2}\left|\langle f|\hat{V}|i\rangle\right|^2 \left[\frac{\sin(\omega_{fi}t/2)}{(\omega_{fi}/2)}\right]^2 . \tag{11.28}$$

Exercise 11.2 For $x \in \mathbb{R}$, show that $|e^{ix} - 1|^2 = 4\sin^2(x/2)$, and use it to derive Eq. (11.28).

The transition probability for constant time t is an oscillating function of ω_{fi}, with a pronounced maximum around the origin, as illustrated in Fig. 11.2. It decreases rapidly as ω_{fi} moves away from the origin, or equivalently, when ω_f departs from ω_i. Note that ω_f can change continuously because the final state $|f\rangle$ is in the continuum. The subsequent peaks are significantly smaller and are separated by zeroes situated at $\omega_{fi} = 2\pi k/t$, where k is an integer. The

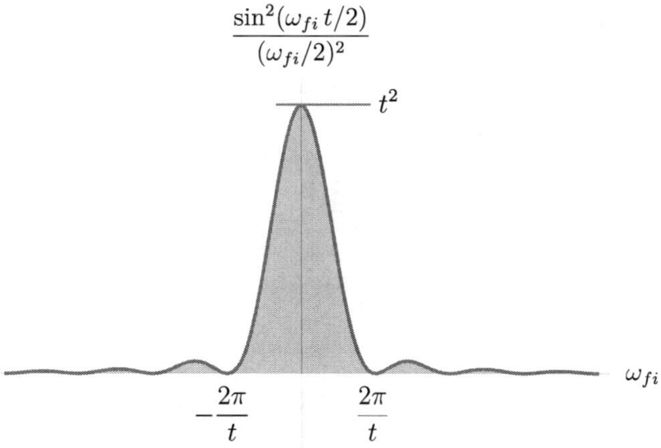

FIGURE 11.2
Plot of the oscillating part of $P_{i\leftarrow i}$ vs. ω_{fi} for a fixed value of t. The zeroes are at $\pm 2\pi/t, \pm 4\pi/t, \cdots$.

central maximum occurs when $\omega_{fi} \simeq 0$ or when the final and initial energies are very close $E_f \simeq E_i$. Because the central fringe is dominant, the largest contribution to the transition probability comes from the interval

$$-\frac{2\pi}{t} \leq \omega_{fi} \leq \frac{2\pi}{t} ,$$

which corresponds to the final energy in the range

$$E_i - \frac{2\pi\hbar}{t} \leq E_f \leq E_i + \frac{2\pi\hbar}{t} . \tag{11.29}$$

The height of the central fringe is proportional to t^2, its width with $1/t$, and thus its area is proportional to t. Consequently, the transition probability increases linearly in time.

At this point we pause for a short mathematical digression. The function

$$f(y,t) = \frac{\sin^2(yt)}{\pi y^2 t}$$

integrates to 1 for all values of t; i.e., $\int_{-\infty}^{\infty} f(y,t)dy = 1$, and its central peak becomes narrower and taller as t increases, as illustrated in Fig. 11.3. It can be shown that in fact

$$\lim_{t\to\infty} \frac{\sin^2(yt)}{\pi y^2 t} = \delta(y) .$$

Taking into account the above observations, for large t we can write

$$\frac{\sin^2(\omega_{fi}t/2)}{(\omega_{fi}/2)^2} \simeq 2\pi t\hbar\, \delta(E_f - E_i) .$$

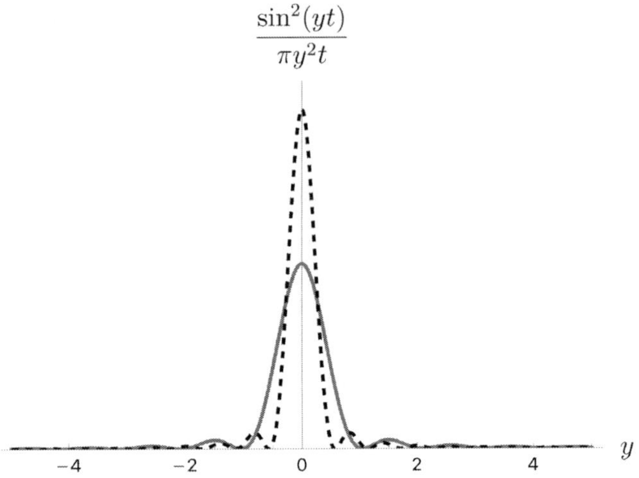

FIGURE 11.3
The function $\sin^2(yt)/(\pi y^2 t)$ for increasing times. The solid line is at $t = 3.0$ and the dashed line at $t = 5.5$. The central peak becomes taller and narrower as t increases, while the area under the curve remains equal to 1. This sequence of functions approaches the δ-Dirac function as $t \to \infty$.

In the above equation, we used the δ-Dirac property $\delta(\omega_{fi}/2) = 2\hbar\delta(\hbar\omega_{fi})$ together with $\hbar\omega_{fi} = \hbar(\omega_f - \omega_i) = E_f - E_i$.

We conclude that the transition probability of Eq. (11.28), in the limit of large t, can be cast in the following form

$$P_{f\leftarrow i}(t) = \frac{2\pi t}{\hbar}\left|\langle f|\hat{V}|i\rangle\right|^2 \delta(E_f - E_i)\ . \tag{11.30}$$

The *transition rate* is defined as the transition probability per unit time, i.e.:

$$\Gamma_{f\leftarrow i} = \frac{P_{f\leftarrow i}(t)}{t} = \frac{2\pi}{\hbar}\left|\langle f|\hat{V}|i\rangle\right|^2 \delta(E_f - E_i)\ . \tag{11.31}$$

The $\delta(E_f - E_i)$ term is in fact an expression of energy conservation. In the large t limit, the transition rate is non-zero only for $E_f \simeq E_i$. We arrived to the conclusion that a constant perturbation does not modify the energy of the system. It just causes transitions which are energy-conserving.

However, the transition rate we obtained contains a δ-Dirac "function" which could be a bit unsettling. Remember that the final state is in the continuum, and as a momentum eigenstate, it is non-normalizable. We will refine further this result by considering the density of the final states $\rho(E_f)$ which gives the number of final states with the energy in the interval between E_f and $E_f + dE_f$:

$$dn(E_f) = \rho(E_f)dE_f\ .$$

We now integrate the transition rate given by Eq. (11.31) over the final states, and obtain the physical result we were looking for:

$$\begin{aligned} W_{f\leftarrow i} &= \int \Gamma_{f\leftarrow i} dn(E_f) = \int \Gamma_{f\leftarrow i}\rho(E_f)dE_f \\ &= \frac{2\pi}{\hbar}\left|\langle f|\hat{V}|i\rangle\right|^2 \int \rho(E_f)\delta(E_f - E_i)dE_f \ , \end{aligned} \tag{11.32}$$

or

$$W_{f\leftarrow i} = \frac{2\pi}{\hbar}\left|\langle f|\hat{V}|i\rangle\right|^2 \rho(E_i) \ , \quad \text{for } E_i \simeq E_f \ . \tag{11.33}$$

This equation is known as *Fermi's Golden Rule*. It tells us that for a constant perturbation, the total rate of transition becomes constant if we wait long enough. [1]

11.5 Harmonic perturbation

We are interested in finding the transition probability in the case of a perturbation that oscillates harmonically in time with angular frequency ω

$$\hat{\mathcal{V}} = \hat{\alpha}e^{i\omega t} + \hat{\alpha}^\dagger e^{-i\omega t} \ , \tag{11.34}$$

where $\hat{\alpha}$ is an "amplitude" operator that does not change in time. Such a perturbation is a good model for the case of the interaction of the electromagnetic radiation with an electron inside the atom. The interaction can stimulate the transition of the electron from one stationary state to another.

We now calculate the transition probability using Eq. (11.23)

$$\begin{aligned} P_{f\leftarrow i}(t) &= \frac{1}{\hbar^2}\left|\langle f|\hat{\alpha}|i\rangle \int_0^t d\tau\, e^{i(\omega_{fi}+\omega)\tau} + \langle f|\hat{\alpha}^\dagger|i\rangle \int_0^t d\tau\, e^{i(\omega_{fi}-\omega)\tau}\right|^2 \\ &= \frac{1}{\hbar^2}\left|\langle f|\hat{\alpha}|i\rangle \frac{e^{i(\omega_{fi}+\omega)t} - 1}{\omega_{fi}+\omega} + \langle f|\hat{\alpha}^\dagger|i\rangle \frac{e^{i(\omega_{fi}-\omega)t} - 1}{\omega_{fi}-\omega}\right|^2 . \end{aligned} \tag{11.35}$$

We have a sum of two complex terms in the probability amplitude. For finite time, each term contains a matrix element ($\langle f|\hat{\alpha}|i\rangle$ or $\langle f|\hat{\alpha}^\dagger|i\rangle$) multiplying a fraction, where the complex number at the numerator is bounded in magnitude to 2. We have two important cases: one when the first term dominates and the other case, when the second term dominates.

[1] Fermi's Golden Rule is a bit of misnomer. It was discovered by Dirac in 1927, more than twenty years before Fermi coined the term. (Dirac, P. A. M. (1 March 1927). "The Quantum Theory of Emission and Absorption of Radiation." Proceedings of the Royal Society A. 114 (767): 243–265). In 1950, Fermi called it "golden rule no. 1" (Fermi, E. (1950). Nuclear Physics. University of Chicago Press. ISBN 978-0226243658, see formula VIII.2). Later on, due to its practical importance it became "Fermi's Golden Rule."

The first term is dominant when $\omega_{fi} + \omega \simeq 0$ or when $E_f \simeq E_i - \hbar\omega$. Since $E_i > E_f$, this happens when the system is initially in an excited state. When acted upon by an external perturbation, the system deexcites and gives up a photon of energy $\hbar\omega$. This is known as the "stimulated emission of radiation." The second term is dominant when $\omega_{fi} - \omega \simeq 0$ or when $E_f \simeq E_i + \hbar\omega$. This

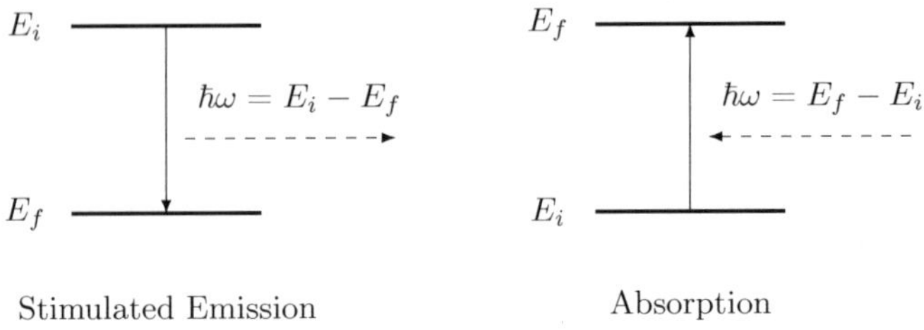

FIGURE 11.4
Stimulated emission and absorption of a photon of energy $\hbar\omega$.

corresponds to the process of "absorption" in which the energy is transferred from the perturbation to the system. The electron absorbs the energy of a photon $\hbar\omega$ and transitions into an excited state. The stimulated emission and absorption of radiation are illustrated in Fig. 11.4.

Both conditions cannot be satisfied at the same time. Thus, in taking the square of the probability amplitude of Eq. (11.35), we can neglect the cross terms, as they contribute negligibly in comparison to the other two. They do not induce lasting transitions. From Eq. (11.35) we thus obtain

$$P_{f\leftarrow i}(t) = \frac{1}{\hbar^2}\left|\langle f|\hat{\alpha}|i\rangle\right|^2 \left|\frac{e^{i(\omega_{fi}+\omega)t}-1}{\omega_{fi}+\omega}\right|^2 + \frac{1}{\hbar^2}\left|\langle f|\hat{\alpha}^\dagger|i\rangle\right|^2 \left|\frac{e^{i(\omega_{fi}-\omega)t}-1}{\omega_{fi}-\omega}\right|^2 .$$

Note that the perturbation in Eq. (11.34) is a Hermitian operator. We have $\langle f|\hat{\alpha}|i\rangle = \langle i|\hat{\alpha}^\dagger|f\rangle$ and consequently $\left|f|\hat{\alpha}|i\right|^2 = \left|\langle f|\hat{\alpha}^\dagger|i\rangle\right|^2$. This yields

$$P_{f\leftarrow i}(t) = \frac{1}{\hbar^2}\left|\langle f|\hat{\alpha}|i\rangle\right|^2 \left(\left[\frac{\sin((\omega_{fi}+\omega)t/2)}{(\omega_{fi}+\omega)/2}\right]^2 + \left[\frac{\sin((\omega_{fi}-\omega)t/2)}{(\omega_{fi}-\omega)/2}\right]^2\right)$$

where as before, we used $|e^{ix}-1|^2 = 4\sin^2(x/2)$. The oscillatory part of the transition probability is illustrated in Fig. 11.5. The two dominant peaks are separated by 2ω. As ω_{fi} moves away from $\pm\omega$ respectively, the transition probability decreases considerably.

In the large t limit, similarly to the previous section, we obtain the transition rate:

$$W_{f\leftarrow i}(t) = \frac{2\pi}{\hbar}\left|\langle f|\hat{\alpha}|i\rangle\right|^2 \Big[\delta(E_f - E_i + \hbar\omega) + \delta(E_f - E_i - \hbar\omega)\Big] \qquad (11.36)$$

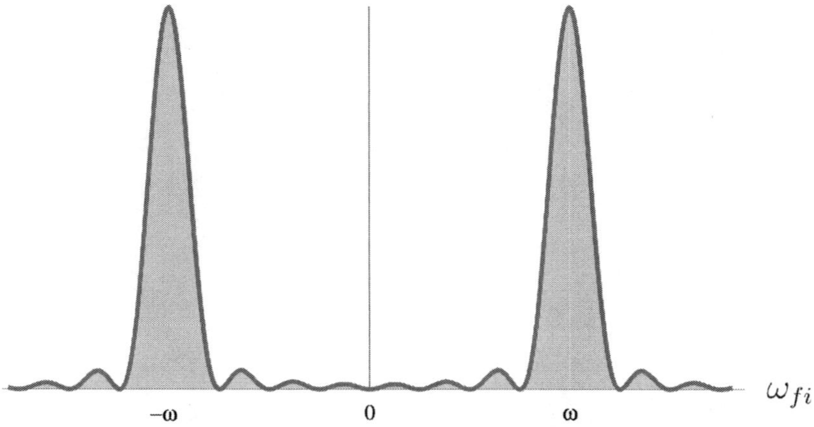

FIGURE 11.5
The oscillatory part of the $P_{f\leftarrow i}$ as a function of ω_{fi}.

This brings to the forefront an important physical observation. The rate is non-zero only when either one of the following two conditions is satisfied:

$$E_f = E_i + \hbar\omega \qquad \text{or} \qquad E_f = E_i - \hbar\omega \ .$$

Clearly, the two conditions cannot be simultaneously true. In the first case, $E_f < E_i$, which indicates that initially the system is in an excited state. In the second case, $E_f > E_i$ and the system ends up into an excited state. The first case thus indeed corresponds to the stimulated emission, while the second case to the absorption of a photon of energy $\hbar\omega$.

Unlike the case of a constant perturbation, which we discussed in the previous section where the energy of the system was conserved, in the case of the harmonic perturbation, the effect of the perturbation is to either remove or inject a photon of energy $\hbar\omega$ into the system.

If the final state is in the continuum, then we can perform a similar analysis with the constant perturbation case described earlier and obtain the emission transition rate

$$W^{\text{emission}}_{f\leftarrow i}(t) = \frac{2\pi}{\hbar} \left| \langle f|\hat{\alpha}|i\rangle \right|^2 \rho(E_f) \ \Big|_{E_f=E_i-\hbar\omega} , \tag{11.37}$$

and the absorption transition rate

$$W^{\text{absorption}}_{f\leftarrow i}(t) = \frac{2\pi}{\hbar} \left| \langle f|\hat{\alpha}^\dagger|i\rangle \right|^2 \rho(E_f) \ \Big|_{E_f=E_i+\hbar\omega} , \tag{11.38}$$

where $\rho(E_f)$ represents the density of states. Because $\left|f|\hat{\alpha}|i\right|^2 = \left|\langle f|\hat{\alpha}^\dagger|i\rangle\right|^2$, from Eqs. (11.37) and (11.37) we get

$$\frac{W_{f\leftarrow i}^{\text{emission}}(t)}{\rho(E_f)\,\big|_{E_f=E_i-\hbar\omega}} = \frac{W_{f\leftarrow i}^{\text{absorption}}(t)}{\rho(E_f)\,\big|_{E_f=E_i+\hbar\omega}} \,. \tag{11.39}$$

This equation is known as the *detailed balancing equation*, and plays an important role in studying the interaction of radiation with mater.

11.6 Problems

11.1 Using Eqs. (11.8) and (11.10), derive Eq. (11.11).

11.2 Prove that:

$$P_{f\leftarrow i}(t) = P_{i\leftarrow f}(t) \,.$$

11.3 A particle is initially in the ground state of an infinite one-dimensional potential well of width L. If at $t = 0$ the particle experiences a perturbation $\hat{\mathcal{V}}(t) = D\,\hat{x}^2 e^{-\beta t}$, calculate the probability of finding the particle in the first excited state. Here D and β are dimensional constants.

12

Scattering

12.1 Introduction

There are several ways of gathering experimental evidence about microscopic systems. Spectral analysis gives information about the bound states of the system, while scattering experiments probe their inner structure. The pioneering alpha scattering experiments by Rutherford yielded the first glimpse into the structure of atoms. Such experiments opened the door to modern collider experiments, and today, using high energy beams, allows us to peer deeper into the structure of matter. In this chapter we will study only elastic collisions, in which the nature of the scattered particles does not change. This is expected for low energies, in the non-relativistic regime. There is another type of collisions, the inelastic collisions, in which particles transition into other particles as a result of the interaction. These are more common at high energies, usually in the relativistic domain and are outside the subject of this book.

We first introduce the concept of cross section and then study two complementary approaches to scattering: the Born approximation, which is suitable when the energy of the incoming particles is high in comparison with the potential energy of the target, and the partial waves method, which applies best when the two energies are comparable.

12.2 Cross section

We consider that the target has a mass much bigger than the mass of the incident particles, and thus it remains at rest in the laboratory frame during the scattering experiment. For example, in the Rutherford experiment, the target was a very thin sheet of gold, and the alpha particles scattered from the atoms inside. We will also consider that the density of scattering centers is such that after scattering, the particle does not experience a secondary scattering with another target particle. Finally, the target that we are probing can be described by a potential confined to a very small volume, in comparison to the macroscopic experimental device. Thus, because the source of the particles and the detector are at macroscopic distances from the target, it is a good

DOI: 10.1201/9781003174967-12

approximation to consider the particle to be a free particle before and after the collision. Therefore, we write the asymptotic wave function of the system as a combination of the incoming plane wave that propagates along the z-axis, and a spherical wave representing the outgoing scattered particle. The target is situated at the origin. If we denote $k = \sqrt{2mE}/\hbar$, we have (see Problem 12.1)

$$\psi(\vec{r})|_{r\to\infty} \sim e^{ikz} + f(\theta,\varphi)\frac{e^{ikr}}{r} = \psi_{\text{inc}}(\vec{r}) + \psi_{\text{out}}(\vec{r}) \ . \tag{12.1}$$

This situation is a bit different from the one-dimensional scattering, where we had only the incident wave and the transmitted and reflected waves.
In three dimensions, the outgoing wave interferes with the incoming wave. The forward scattered wave, which we can interpret as the transmitted wave, would be diminished as result of this interference. Because the total flux of particles is conserved, this is essential for having scattered particles at angles θ away from the z-axis.

If the solid angle $d\Omega$ represents the angle subtended by the detector, then we define the *differential cross section* $d\sigma/d\Omega$ as the ratio of the number of particles n_{det} scattered into $d\Omega$ per unit time and the j_{inc}, which is the number of particles incident on the target per unit area per unit time. We thus have

$$\frac{d\sigma}{d\Omega} = \frac{n_{\text{det}}}{j_{\text{inc}}} \ . \tag{12.2}$$

Notice that $d\sigma/d\Omega$ has units of area. If we now integrate over all angles, we obtain the total cross section:

$$\sigma = \int \frac{d\sigma}{d\Omega} d\Omega \ . \tag{12.3}$$

To connect the flux of particles with the probability current, we start with the Schrödinger equation in position space

$$-\frac{\hbar^2}{2m}\nabla^2\psi(\vec{r}) + V(\vec{r})\,\psi(\vec{r}) = E\,\psi(\vec{r}) \tag{12.4}$$

and calculate the time derivative of the probability density $\rho = \psi^*\psi$. We obtain the continuity equation

$$\frac{\partial\rho}{\partial t} + \vec{\nabla}\cdot\vec{j} = 0 \ , \tag{12.5}$$

where $\vec{j}$ is the probability current:

$$\vec{j} = \frac{i\hbar}{2m}\left(\psi\vec{\nabla}\psi^* - \psi^*\vec{\nabla}\psi\right) \ . \tag{12.6}$$

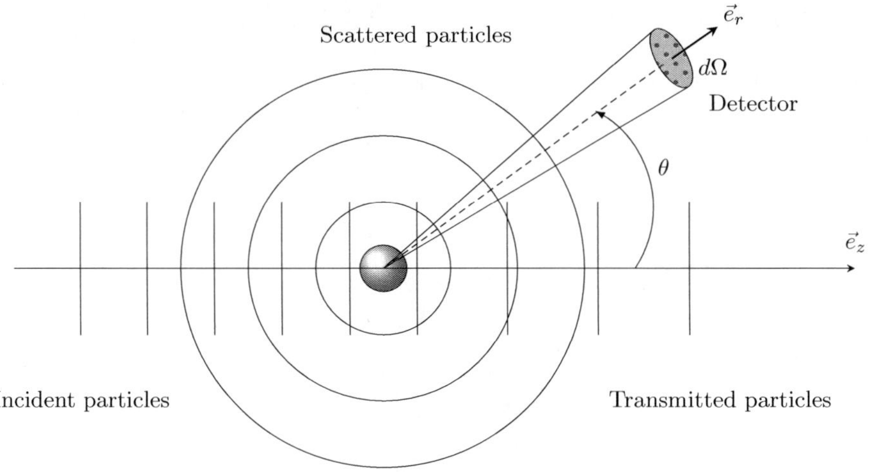

FIGURE 12.1
Illustration of the asymptotic representation of wave functions during an elastic scattering experiment.

Exercise 12.1 Show that the probability current has dimensions of (Area × Time)$^{-1}$.

In practice, we consider the number of particles entering a detector that subtends a solid angle $d\Omega$, which is away at an angle θ from the forward direction z, as illustrated in Fig. 12.1. The flux of scattered particles is proportional to the outgoing probability current $\vec{j}_{\text{out}}$:

$$\vec{j}_{\text{out}} = \frac{i\hbar}{2m}\left(\psi_{\text{out}}\vec{\nabla}\psi^*_{\text{out}} - \psi^*_{\text{out}}\vec{\nabla}\psi_{\text{out}}\right) , \tag{12.7}$$

where

$$\psi_{\text{out}} = f(\theta,\varphi)\frac{e^{ikr}}{r} .$$

Note that, in the spherical coordinates with unit vectors $\{\vec{e}_r, \vec{e}_\theta, \vec{e}_\varphi\}$ the gradient of ψ_{out} in the asymptotic limit $r \to \infty$ gives

$$\vec{\nabla}\psi_{\text{out}} = \left(\frac{ik}{r}e^{ikr} - \frac{1}{r^2}\right) f\,\vec{e}_r + \frac{1}{r}\frac{\partial f}{\partial\theta}\frac{ik}{r}e^{ikr}\vec{e}_\theta + \frac{1}{r\sin\theta}\frac{\partial f}{\partial\varphi}\frac{ik}{r}e^{ikr}\vec{e}_\varphi \simeq f\frac{ik}{r}e^{ikr}\vec{e}_r$$

In the last step we discarded all terms decreasing like $1/r^2$. Then, the outgoing probability current becomes

$$\vec{j}_{\text{out}} = \frac{\hbar k}{mr^2}|f(\theta,\varphi)|^2\,\vec{e}_r . \tag{12.8}$$

Note that, as expected, this current is in the radial direction $\vec{e}_r$.

Exercise 12.2 Derive Eq. (12.8).

The outgoing probability flow, which is proportional with the flux of scattered particles, is obtained by taking the dot product of the outgoing current with the infinitesimal area of the detector subtending the solid angle $d\Omega$

$$n_{det} = \vec{j}_{\text{out}} \cdot (r^2 d\Omega)\, \vec{e}_r = \frac{\hbar k}{m} |f(\theta,\varphi)|^2\, d\Omega \ . \tag{12.9}$$

Similarly, the incident current is given by

$$\vec{j}_{\text{inc}} = \frac{i\hbar}{2m}\left(\psi_{\text{inc}} \vec{\nabla}\psi^*_{\text{inc}} - \psi^*_{\text{inc}}\vec{\nabla}\psi_{\text{inc}}\right) = \frac{\hbar k}{m}\,\vec{e}_z \ , \tag{12.10}$$

where

$$\psi_{in} = e^{ikz} \ .$$

The incident flux of particles on the target j_{inc} is

$$j_{\text{inc}} = \vec{j}_{\text{inc}} \cdot \vec{e}_z = \frac{\hbar k}{m} \ . \tag{12.11}$$

Finally, substituting Eqs. (12.9) and (12.11) into (12.2), we obtain

$$\frac{d\sigma}{d\Omega} = |f(\theta,\varphi)|^2 \ . \tag{12.12}$$

The function $f(\theta,\varphi)$ is known as the *scattering amplitude.*

12.3 Born approximation

Born approximation applies best in the case in which the energy E of the incident beam is much larger than the magnitude of the potential $V(\vec{r})$, which we consider confined to a small volume around the target particle. With the usual notation $k^2 = 2mE/\hbar^2$, we rewrite Schrödinger equation (12.4) as

$$(\vec{\nabla}^2 + k^2)\,\psi = \frac{2m}{\hbar^2}\, V\psi \ . \tag{12.13}$$

To solve this equation, in the first step we look for a solution of the homogeneous equation

$$(\vec{\nabla}^2 + k^2)\,\psi = 0 \ , \tag{12.14}$$

and find that incoming wave $\psi_{\text{inc}} = e^{ikz}$ is such a solution. In the next step, we look for the Green's function $G(\vec{r},\vec{r}\,')$ associated with Eq. (12.4); i.e.,

$$(\vec{\nabla}^2 + k^2)\, G(\vec{r},\vec{r}\,') = \delta^3(\vec{r} - \vec{r}\,') \ . \tag{12.15}$$

Assuming that we found the Green's function, we write the formal solution of the Schrödinger equation (12.13) as

$$\psi(\vec{r}) = e^{ikz} + \frac{2m}{\hbar^2}\int \mathrm{d}^3r'\, G(\vec{r},\vec{r}\,')\, V(\vec{r}\,')\, \psi(\vec{r}\,') \,. \tag{12.16}$$

Exercise 12.3 Applying $(\vec{\nabla}^2 + k^2)$ on Eq. (12.16) show that it satisfies Eq. (12.13).

Because ψ appears on both sides of Eq. (12.16), this is not a "true" solution. However, it can be used in obtaining an iterative solution as following: first, we find the Green's function $G(\vec{r},\vec{r}\,')$. Then, as a first approximation, we replace the ψ under the integral with the incident wave e^{ikz}. Indeed, if $V = 0$ then there would be no scattering, and the plane wave would pass undisturbed. In our case, the replacement is justified because V is small in comparison with E.

This procedure is at the core of Born approximation. It can be iterated to calculate the next order approximation. The zero-th order approximation is

$$\psi^{(0)}(\vec{r}) = e^{ikz} \,. \tag{12.17}$$

We use the zero-th order to calculate the first order

$$\psi^{(1)}(\vec{r}) = \psi^{(0)}(\vec{r}) + \frac{2m}{\hbar^2}\int \mathrm{d}^3r'\, G(\vec{r},\vec{r}\,')\, V(\vec{r}\,')\, \psi^{(0)}(\vec{r}\,') \,, \tag{12.18}$$

which we substitute back into the integral to calculate the next order, and so on. We obtain:

$$\begin{aligned}\psi(\vec{r}) = {} & \psi^{(0)}(\vec{r}) + \frac{2m}{\hbar^2}\int \mathrm{d}^3r'\, G(\vec{r},\vec{r}\,')\, V(\vec{r}\,')\, \psi^{(0)}(\vec{r}\,') \\ & + \left(\frac{2m}{\hbar^2}\right)^2 \int \mathrm{d}^3r' \int \mathrm{d}^3r''\, G(\vec{r},\vec{r}\,')\, V(\vec{r}\,') G(\vec{r}\,',\vec{r}\,'')\, V(\vec{r}\,'')\, \psi^{(0)}(\vec{r}\,'') \\ & + \cdots \end{aligned} \tag{12.19}$$

Having established the procedure, we next find the Green's function of Eq.(12.15). To simplify the calculation, because G depends on the difference $\vec{r} - \vec{r}\,'$, in the first step we set $\vec{r}\,' = 0$. At the end we reinstate $\vec{r}\,'$ to obtain the general Green's function. We have

$$\left(\vec{\nabla}^2 + k^2\right) G(\vec{r},0) = \delta^3(\vec{r}) \,. \tag{12.20}$$

For $r \neq 0$ the function $\delta(\vec{r}) = 0$ and Ae^{ikr}/r is a solution of Eq. (12.20), so we explore whether this could be the sought after Green's function. We therefore try the following ansatz

$$G(\vec{r},0) = A\frac{e^{ikr}}{r} \,, \tag{12.21}$$

where A is a constant. Due to the spherical symmetry of the ansatz, it makes sense to use spherical coordinates. Recall that in spherical coordinates the gradient is

$$\vec{\nabla} = \vec{e}_r \frac{\partial}{\partial r} + \vec{e}_\theta \frac{1}{r}\frac{\partial}{\partial \theta} + \vec{e}_\varphi \frac{1}{r\sin\theta}\frac{\partial}{\partial \varphi} .$$

To check whether Eq. (12.21) is a solution of (12.20) for a given A, we first calculate:

$$\begin{aligned}
\vec{\nabla}^2 \frac{e^{ikr}}{r} &= \vec{\nabla}\cdot\left(\vec{\nabla}\frac{e^{ikr}}{r}\right) = \vec{\nabla}\cdot\left(\frac{1}{r}\vec{\nabla}e^{ikr} + e^{ikr}\vec{\nabla}\frac{1}{r}\right) \\
&= 2\vec{\nabla}\left(\frac{1}{r}\right)\cdot\vec{\nabla}\left(e^{ikr}\right) + \frac{1}{r}\vec{\nabla}^2\left(e^{ikr}\right) + e^{ikr}\vec{\nabla}^2\left(\frac{1}{r}\right) \\
&= -k^2\frac{e^{ikr}}{r} + e^{ikr}\vec{\nabla}^2\left(\frac{1}{r}\right) \\
&= -k^2\frac{e^{ikr}}{r} - 4\pi e^{ikr}\delta^3(\vec{r}) . \qquad (12.22)
\end{aligned}$$

In deriving the last line of Eq. (12.22) we used the identity

$$\vec{\nabla}^2\left(\frac{1}{r}\right) = -4\pi\delta^3(\vec{r}) , \qquad (12.23)$$

which can be proven by integrating it over an arbitrary spherical volume $\mathcal{V}$ enclosing the origin. Applying the Gauss's theorem to the left-hand side, we obtain

$$LHS = \int_{\mathcal{V}} \mathrm{d}^3r\, \vec{\nabla}\cdot\vec{\nabla}\left(\frac{1}{r}\right) = \oint_{\Sigma} \vec{e}_r\cdot\vec{\nabla}\left(\frac{1}{r}\right)\mathrm{d}S = \oint_{\Sigma}\left(-\frac{1}{r^2}\right)r^2\mathrm{d}\Omega = -4\pi .$$

The right-hand side gives

$$RHS = \int_{\mathcal{V}} \mathrm{d}^3r\, \left(-4\pi\,\delta^3(\vec{r})\right) = -4\pi ,$$

which completes the proof. We therefore obtain

$$\left(\vec{\nabla}^2 + k^2\right)\left(-\frac{1}{4\pi}\frac{e^{ikr}}{r}\right) = \delta^3(\vec{r}) , \qquad (12.24)$$

and the Green's function is:

$$G(\vec{r},0) = -\frac{1}{4\pi}\frac{e^{ikr}}{r} . \qquad (12.25)$$

Reinstating $\vec{r}\,'$, we obtain the general form of the Green's function associated with the Schrödinger equation (12.13)

$$G(\vec{r},\vec{r}\,') = -\frac{1}{4\pi}\frac{e^{ik|\vec{r}-\vec{r}\,'|}}{|\vec{r}-\vec{r}\,'|} , \qquad (12.26)$$

and consequently, its formal solution:

$$\psi(\vec{r}) = e^{ikz} - \frac{m}{2\pi\hbar^2} \int d^3r' \, \frac{e^{ik|\vec{r}-\vec{r}\,'|}}{|\vec{r}-\vec{r}\,'|} \, V(\vec{r}\,')\, \psi(\vec{r}\,') \, . \tag{12.27}$$

The detector is located at a point far from the scattering center, thus $|\vec{r}| \gg |\vec{r}\,'|$ as shown in Fig. 12.2. For the next step in our approximation we evaluate Green's function for large r. Because $|\vec{r}| \gg |\vec{r}\,'|$, we approximate $1/|\vec{r}-\vec{r}\,'|$ with $1/r$. For the complex exponential term, since it is an oscillatory function, we retain the first order term in r'/r. We have $|\vec{r}-\vec{r}\,'| \simeq r\,(1 - 2\vec{e}_r \cdot \vec{r}\,'/r)^{1/2} \simeq r\,(1 - \vec{e}_r \cdot \vec{r}\,'/r)$. Consequently, for large r we can write

$$\frac{e^{ik|\vec{r}-\vec{r}\,'|}}{|\vec{r}-\vec{r}\,'|} \simeq \frac{1}{r}\, e^{ikr} e^{-ik\vec{e}_r\cdot\vec{r}\,'} = \frac{1}{r}\, e^{ikr} e^{-i\vec{k}_r\cdot\vec{r}\,'} \, , \tag{12.28}$$

where $\vec{e}_r$ is the unit vector in $\vec{r}$ direction, and $\vec{k}_r = k\vec{e}_r$ is the wave number of the scattered wave which is pointing in the direction of $\vec{r}$.

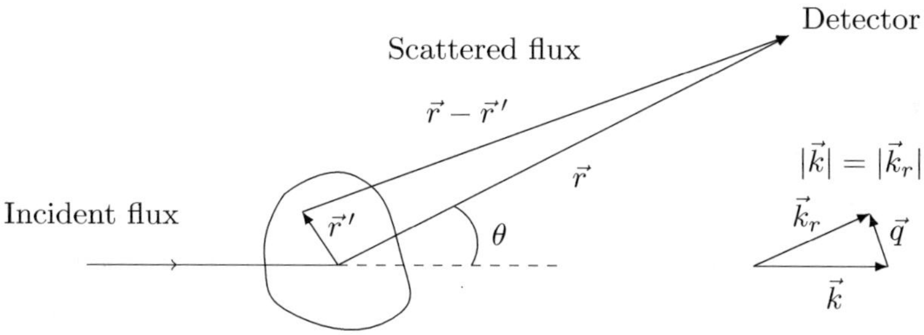

FIGURE 12.2
The typical range of the interaction for the potential V is very small. This effectively restricts the range of r', and thus r' is very small in comparison with r, where the detector is placed.

Putting together all of these back into Eq. (12.27), we obtain in the large r limit

$$\psi(\vec{r}) = e^{ikz} - \frac{m}{2\pi\hbar^2} \frac{e^{ikr}}{r} \int d^3r' \, e^{-i\vec{k}_r\cdot\vec{r}\,'} \, V(\vec{r}\,')\, \psi(\vec{r}\,') \, . \tag{12.29}$$

We now have all the elements needed to discuss Born approximation. Because the energy of the incoming particles is much larger than the potential V of the target, in the first approximation we replace $\psi(r')$ in the integral of Eq. (12.29) with the plane wave modeling the incident flux of particles. We obtain the first Born approximation:

$$\psi(\vec{r}) = e^{ikz} - \frac{m}{2\pi\hbar^2} \frac{e^{ikr}}{r} \int d^3r' \, e^{-i\vec{k}_r\cdot\vec{r}\,'} \, V(\vec{r}\,')\, e^{ikz'} \, . \tag{12.30}$$

This expression can be further refined by noticing that $kz' = \vec{k}\cdot\vec{r}\,'$, where $\vec{k}$ is the wave number of the incoming particles. Then, the exponential terms give $-i(\vec{k}_r - \vec{k})\cdot\vec{r}\,' = -i\vec{q}\cdot\vec{r}\,'$. Here, $\vec{q}$ is connected to the change in the momentum of the particle due to the collision $\hbar\,\vec{q} = \hbar(\vec{k}_r - \vec{k}) = \vec{p}_r - \vec{p}$. We obtain finally

$$\psi(\vec{r}) = e^{ikz} - \frac{m}{2\pi\hbar^2}\frac{e^{ikr}}{r}\int \mathrm{d}^3r'\; V(\vec{r}\,')\, e^{-i\vec{q}\cdot\vec{r}\,'} \;. \tag{12.31}$$

Comparing this result with Eq. (12.1), we see that the scattering amplitude in the first Born approximation can be written as

$$f(\theta,\varphi) = -\frac{m}{2\pi\hbar^2}\int \mathrm{d}^3r\; V(\vec{r})\, e^{-i\vec{q}\cdot\vec{r}} \;, \tag{12.32}$$

where for simplicity, we dropped the prime sign from the integration variable. Note that in this order of approximation, the scattering amplitude is proportional to the Fourier transform of the potential.

12.3.1 Central potentials

In the case of central potentials the calculation of the scattering amplitude simplifies. Because we chose the z-axis as the incident direction, the scattering angle θ is also the polar angle of the coordinate system, as illustrated in Fig. 12.2. We obtain

$$q^2 = (\vec{k}_r - \vec{k})\cdot(\vec{k}_r - \vec{k}) = 2k^2 - 2k^2\cos\theta = 4k^2\sin^2(\theta/2)\;.$$

Thus, we write the magnitude of $\vec{q}$ in terms of the polar angle θ, and the wave number $k = \sqrt{2mE}/\hbar$:

$$q = 2k\sin(\theta/2), \qquad 0 \le \theta \le \pi\;. \tag{12.33}$$

As expected, the incident particle experiences the maximum change in momentum when $\theta = \pi$ or when it is scattered backward. Because $\vec{q}\cdot\vec{r} = qr\cos\theta$ and $V(\vec{r}) = V(r)$ in Eq. (12.32), we denote $f(\theta,\varphi) = f(\theta) \equiv f(q)$ and obtain

$$\begin{aligned} f(q) &= -\frac{m}{2\pi\hbar^2}\int_0^\infty V(r)\,r^2dr\int_0^{2\pi}d\varphi\int_0^\pi d\theta\,\sin\theta\, e^{-iqr\,\cos\theta} \\ &= -\frac{m}{2\pi\hbar^2}\frac{2\pi}{iq}\int_0^\infty rV(r)\left(e^{iqr} - e^{-iqr}\right)dr \\ &= -\frac{2m}{\hbar^2 q}\int_0^\infty rV(r)\sin(qr)\,dr\;, \end{aligned} \tag{12.34}$$

or equivalently

$$f(\theta) = -\frac{2m}{\hbar^2 q}\int_0^\infty rV(r)\sin(qr)\,dr\;, \quad q = 2k\sin(\theta/2)\;. \tag{12.35}$$

Note that for a rapidly decreasing potential, the integral produces a finite result. However for a $1/r$ potential, such as the Coulomb potential, the integral is ill defined.

Example 12.1 Scattering from Yukawa potential
As an example of scattering from a central potential we choose the modified Coulomb potential

$$V_\lambda(r) = \frac{K}{r} e^{-\lambda r} \, . \tag{12.36}$$

Here, we introduced the exponential factor to mitigate the convergence issue caused by the Coulomb potential's $1/r$ dependency. This is a screened Coulomb potential and was initially introduced by Yukawa to model the strong interactions between nucleons. When $\lambda \to 0$, for $K = Z_1 Z_2 e^2$, we recover the Coulomb potential describing the interaction between an incident particle of charge $Z_1 e$ and a nucleus of charge $Z_2 e$.

The Born approximation gives

$$\begin{aligned} f_\lambda(q) &= -\frac{2m}{\hbar^2}\frac{K}{q}\int_0^\infty e^{-\lambda r}\sin(qr)\,dr \\ &= -\frac{2m}{\hbar^2}\frac{K}{q^2+\lambda^2} \, . \end{aligned} \tag{12.37}$$

Exercise 12.4 Prove Eq. (12.37).

Taking now $\lambda \to 0$, we obtain the scattering amplitude for the Coulomb potential

$$f(q) = \lim_{\lambda\to 0} f_\lambda(q) = -\frac{2m}{\hbar^2}\frac{K}{q^2} \, , \tag{12.38}$$

that yields the differential cross section

$$\frac{d\sigma}{d\Omega} = \left(\frac{2m}{\hbar^2}\frac{K}{q^2}\right)^2 \, . \tag{12.39}$$

Substituting back $q = 2k\sin(\theta/2)$ and $k = \sqrt{2mE}/\hbar$ in Eq. (12.38), we finally obtain

$$\frac{d\sigma}{d\Omega} = \left(\frac{K}{4E}\right)^2 \frac{1}{\sin^4\theta/2} \, , \tag{12.40}$$

which, surprisingly, is identical to the classical Rutherford cross section formula. Such a result can be understood if we remember that the Born approximation is valid in the high-energy limit, for which classical and quantum results typically agree.

12.4 Partial wave analysis

The partial wave method can be applied to central potentials of finite range, i.e., potentials for which $V(r) \simeq 0$, for $r > r_0$. In contrast with the Born approximation, which applies primarily to relatively high energies, the partial wave analysis is best suited to low energies, when the energy of the incoming particles is comparable to the potential energy of the target. The key element of the method is the conservation of the angular momentum in central fields. Thus, the incident particles of a given angular momentum can be analyzed independently of the particles of another angular momentum.

12.4.1 Partial waves

Remember that the incident beam was considered to be a plane wave of definite linear momentum $\vec{p} = \hbar\vec{k}$. However, the angular momentum can have arbitrary values. The plane wave has an infinite extension, corresponding to a classical impact parameter b varying from zero to infinity.

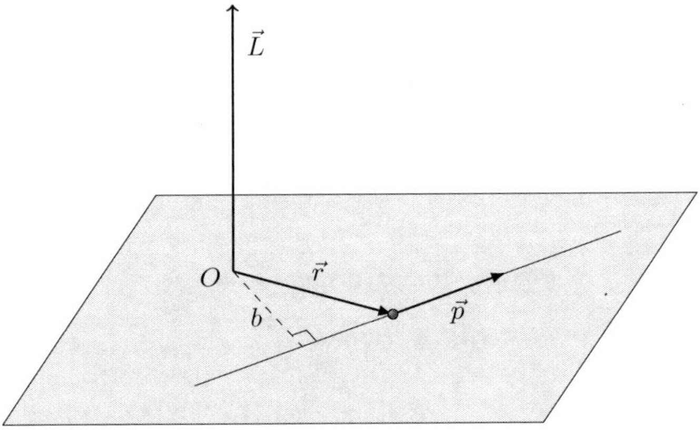

FIGURE 12.3
The classical impact parameter b of a particle of momentum $\vec{p}$ and angular momentum $\vec{L}$ relative to the target placed at the origin O.

Thus, a plane wave carries angular momenta ranging from zero to infinity. We can in principle, analyze a plane wave independently for each angular momentum. We call such a component a *partial wave*, and the process of decomposing a plane wave into partial waves, *partial wave analysis*. In practice, only the first few components contribute significantly to the total cross-section, as we will see shortly.

As we have done in the previous section, we will consider that the incident wave propagates along the z-axis. Due to the spherical symmetry of the potential, the scattering amplitude $f(\theta, \varphi)$ will not depend on the azimuthal angle φ. Then, both the incident wave

$$\psi_{\text{inc}}(\vec{r}) \sim e^{ikz} = e^{ikr\cos\theta} , \tag{12.41}$$

and the total wave function for large r

$$\psi(\vec{r}) \sim e^{ikr\cos\theta} + f(\theta)\frac{e^{ikr}}{r} , \tag{12.42}$$

will depend only on the axial angle θ. Because the action of the central potential $V(r)$ has a limited range, outside this range, both wave functions satisfy the Schrödinger equation for the free particle

$$\vec{\nabla}^2\psi(\vec{r}) + k^2\psi(\vec{r}) = 0 . \tag{12.43}$$

The most general solution in spherical coordinates can be written in terms of the radial function $R_l(r)$ and the spherical harmonics $Y_{l,m}(\theta,\varphi)$ [1]. But due to the spherical symmetry, the term $e^{im\varphi}$ must disappear because the solution doesn't depend on φ. Hence $m = 0$, and we obtain the wave function satisfying Eq. (12.43) for the spherically symmetric case

$$\psi(r,\theta) = \sum_{l=0}^{\infty} R_l(r)\, P_l(\cos\theta) , \tag{12.44}$$

where $P_l(\cos\theta) = \sqrt{\frac{4\pi}{2l+1}}Y_{l,0}(\theta,\varphi)$ are the Legendre polynomials of order l. The Legendre polynomials $P_l(\cos\theta)$ form a complete set in the θ-space and thus the products $R_l(r)\,P_l(\cos\theta)$ span the entire r,θ-space. Substituting Eq. (12.44) into (12.43) gives the radial equation

$$r^2\frac{d^2R_l}{dr^2} + 2r\frac{dR_l}{dr} + \left[k^2r^2 - l(l+1)\right]R_l = 0 , \tag{12.45}$$

whose solution is a linear combination of the spherical Bessel functions $j_l(kr)$ and $y_l(kr)$. It is convenient to write these functions using Rayleigh's formula

$$j_n(x) = (-x)^n\left(\frac{1}{x}\frac{d}{dx}\right)^n\frac{\sin x}{x} \quad ; \quad y_n(x) = -(-x)^n\left(\frac{1}{x}\frac{d}{dx}\right)^n\frac{\cos x}{x} ,$$

which reveal the behaviors of $j_n(x)$ and $y_n(x)$ near origin. Specifically, the functions $j_n(x)$ are finite, whereas $y_n(x)$ become singular, as illustrated in Fig. 12.4. The asymptotic behavior of these functions in the limit $x \to \infty$ is

$$j_n(x) \to \frac{\sin(x - n\pi/2)}{x} \quad ; \quad y_n(x) \to -\frac{\cos(x - n\pi/2)}{x} . \tag{12.46}$$

which indicates that they are both well-behaved far away from the origin.

[1] Where, $Y_{l,m}(\theta,\varphi) = \frac{(-1)^l}{2^l\, l!}\sqrt{\frac{(2l+1)(l+m)!}{4\pi(l-m)!}}e^{im\varphi}\frac{1}{\sin^m\theta}\frac{d^{l-m}}{d(\cos\theta)^{l-m}}\sin^{2l}\theta$.

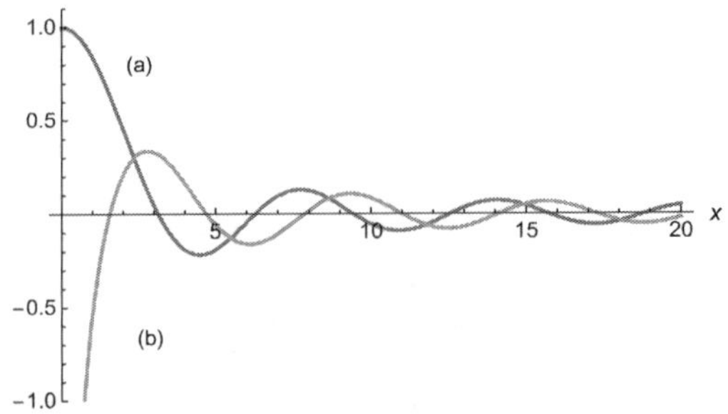

FIGURE 12.4
Spherical Bessel functions: (a) $j_0(x)$ and (b) $y_0(x)$. Note that at $x = 0$, $y_0(x)$ becomes singular.

12.4.2 Free wave expansion

With these preliminaries we now return to the expression for the incident wave Eq. (12.41), which we write with the help of Eq. (12.44) as

$$\psi_{\text{inc}}(r,\theta) \sim e^{ikr\cos\theta} = \sum_{l=0}^{\infty} A_l\, j_l(kr)\, P_l(\cos\theta) \;, \tag{12.47}$$

where A_l are constants. Note that there are no $y_l(kr)$ in this expression because they diverge at $r \to 0$. To find the constants A_l we use the orthogonality of Legendre polynomials

$$\int_{-1}^{1} P_l(x)P_{l'}(x)dx = \frac{2\,\delta_{ll'}}{2l+1} \qquad l, l' = 0, 1, 2, \ldots \tag{12.48}$$

and obtain

$$A_l\, j_l(kr) = (l + 1/2)\int_{-1}^{1} e^{ikrx} P_l(x)dx \;. \tag{12.49}$$

> **Exercise 12.5** Use the orthogonality of the Legendre polynomials to prove Eq. (12.49) by judiciously integrating Eq. (12.47).

We know [2] that the spherical Bessel function $j_l(kr)$ can also be written as:

$$j_l(kr) = \frac{(-i)^l}{2}\int_{-1}^{1} e^{ikrx} P_l(x)dx \;. \tag{12.50}$$

Comparing now Eq. (12.50) with (12.49), we obtain the expression for the coefficients A_l

$$A_l = i^l(2l+1) \;, \tag{12.51}$$

[2]See for example Abramowitz and Stegun, (Dover NY, 1965) Eq. 10.1.14

which give the decomposition of the incident plane wave into a series of spherical waves

$$\psi_{\text{inc}}(r,\theta) \sim e^{ikr\cos\theta} = \sum_{l=0}^{\infty} i^l (2l+1)\, j_l(kr)\, P_l(\cos\theta)\ . \qquad (12.52)$$

Each spherical component corresponds to a given angular momentum and is called a *partial wave*. The asymptotic form of Eq. (12.52) becomes

$$\psi_{\text{inc}}(r,\theta) \to e^{ikr\cos\theta} = \sum_{l=0}^{\infty} i^l (2l+1)\, \frac{1}{kr} \sin(kr - l\pi/2)\, P_l(\cos\theta)\ . \qquad (12.53)$$

12.4.3 Scattered waves and phase shifts

Similarly to Eq. (12.52), far from the target region, the solution of the Schrödinger equation for the scattered wave can be written as

$$\psi(r,\theta) \sim \sum_{l=0}^{\infty} i^l (2l+1)\, R_l(r)\, P_l(\cos\theta)\ . \qquad (12.54)$$

However, the radial part, which as we will see later, encodes the effects of the scattering potential, can now be written as a linear combination of $j_l(kr)$ and $y_l(kr)$, because for $r \to \infty$ they are both well-behaved.

$$\psi(r,\theta) \sim \sum_{l=0}^{\infty} i^l (2l+1) \left[B_l\, j_l(kr) + C_l\, y_l(kr)\right] P_l(\cos\theta)\ . \qquad (12.55)$$

Using the asymptotic forms for $j_l(kr)$ and $y_l(kr)$ we write

$$\begin{aligned} \psi(r,\theta) \quad &\to \quad \sum_{l=0}^{\infty} i^l (2l+1) \left[B_l\, \frac{\sin(kr - l\pi/2)}{kr} - C_l\, \frac{\cos(kr - l\pi/2)}{kr}\right] P_l(\cos\theta) \\ &= \quad \sum_{l=0}^{\infty} i^l (2l+1) \frac{D_l}{kr} \sin(kr - l\pi/2 + \delta_l)\, P_l(\cos\theta)\ , \end{aligned} \qquad (12.56)$$

where we reparameterized the constants B_l and C_l as $B_l = D_l \cos\delta_l$ and $C_l = -D_l \sin\delta_l$, respectively. The l dependent phase term δ_l is called a *phase shift*.

At this point we will find the coefficients D_l by fitting the wave $\psi(r,\theta)$ of Eq. (12.56) to the asymptotic solution of the Schrödinger equation

$$\psi(r,\theta) \sim e^{ikr\cos\theta} + f(\theta)\frac{e^{ikr}}{r}\ ,$$

where $e^{ikr\cos\theta}$ is given by Eq. (12.53). We obtain

$$\begin{aligned} f(\theta)e^{ikr} &= r\,\psi(r,\theta) - r\,e^{ikr\cos\theta} \\ &= \sum_{l=0}^{\infty} i^l(2l+1)\frac{D_l}{k}\sin(kr - l\pi/2 + \delta_l)\,P_l(\cos\theta) \\ &- \sum_{l=0}^{\infty} i^l(2l+1)\,\frac{1}{k}\sin(kr - l\pi/2)\,P_l(\cos\theta)\;. \end{aligned} \tag{12.57}$$

In the next step we use Euler's formula, $\sin x = (e^{ix} - e^{-ix})/2i$, and factor out the common terms:

$$\begin{aligned} f(\theta)e^{ikr} &= \left\{\sum_{l=0}^{\infty} i^l(2l+1)\,P_l(\cos\theta)\frac{e^{-il\pi/2}}{2ik}\left[D_l e^{i\delta_l} - 1\right]\right\} e^{ikr} \quad (12.58) \\ &- \left\{\sum_{l=0}^{\infty} i^l(2l+1)\,P_l(\cos\theta)\frac{e^{il\pi/2}}{2ik}\left[D_l e^{-i\delta_l} - 1\right]\right\} e^{-ikr}\;. \end{aligned}$$

Exercise 12.6 Derive Eq. (12.58).

We see now that in Eq. (12.58) the coefficient of e^{-ikr} must vanish, whilst the coefficient of e^{ikr} must give $f(\theta)$. We thus obtain

$$D_l = e^{i\delta_l}\;, \tag{12.59}$$

and

$$f(\theta) = \frac{1}{2ik}\sum_{l=0}^{\infty}(2l+1)\,P_l(\cos\theta)\left(e^{2i\delta_l} - 1\right) \tag{12.60}$$

or

$$f(\theta) = \frac{1}{k}\sum_{l=0}^{\infty}(2l+1)\,e^{i\delta_l}\sin\delta_l P_l(\cos\theta)\;. \tag{12.61}$$

Eq. (12.61) is the partial wave expansion of the scattering amplitude which we will use to calculate the differential cross-section $d\sigma/d\Omega = |f(\theta)|^2$ and then the total cross-section σ. We have

$$\begin{aligned} \sigma &= \int \frac{d\sigma}{d\Omega}\,d\Omega = 2\pi\int_{-1}^{1}|f(\theta)|^2 d(\cos\theta) \\ &= \frac{2\pi}{k^2}\sum_{l,l'=0}^{\infty}(2l+1)(2l'+1)\,e^{i(\delta_l-\delta_{l'})}\sin\delta_l\sin\delta_{l'}\int_{-1}^{1}P_l(x)P_{l'}(x)\,dx \\ &= \frac{4\pi}{k^2}\sum_{l=0}^{\infty}(2l+1)\sin^2\delta_l = \sum_{l=0}^{\infty}\sigma_l\;, \end{aligned} \tag{12.62}$$

where we used the orthogonality of the Legendre polynomials Eq. (12.48). The quantities σ_l are known as *partial wave cross-sections.* This decomposition in l justifies the breaking down of our analysis in independent partial waves for central potentials.

12.4.4 The optical theorem

If in Eq. (12.61) we choose $\theta = 0$, taking into account that $P_l(1) = 1$, we obtain the expression for the forward scattering amplitude:

$$f(0) = \frac{1}{k}\sum_{l=0}^{\infty}(2l+1)\,e^{i\delta_l}\sin\delta_l = \sum_{l=0}^{\infty}(2l+1)\,\left(\cos\delta_l\sin\delta_l + i\sin^2\delta_l\right)\ . \tag{12.63}$$

The imaginary part of $f(0)$ is then

$$\mathrm{Im}\,f(0) = \sum_{l=0}^{\infty}(2l+1)\,\sin^2\delta_l = \frac{k}{4\pi}\sigma\ , \tag{12.64}$$

from which we obtain

$$\sigma = \frac{4\pi}{k}\mathrm{Im}\,f(0)\ . \tag{12.65}$$

We can connect the total cross-section with the loss of intensity suffered by the incident beam in the forward direction, i.e., at $\theta = 0$. This is because some particles have been deflected away from the incident direction. This result is called the *Optical Theorem* in analogy with a similar case in optics where the imaginary part of the complex reflection index is related to the absorption cross-section for light in that medium.

12.4.5 The dependence of δ_l on $V(r)$

The influence of the potential is reflected by the phase shifts δ_l in the wave function. When the potential is zero, obviously, $\delta_l = 0$. If we make the change of variable $u_l(r) = rR_l(r)$ the radial equation Eq. (12.45) becomes:

$$\left[\frac{d^2}{dr^2} + k^2 - \left(\frac{2m}{\hbar^2}V(r) + \frac{l(l+1)}{r^2}\right)\right]\,u_l(r) = 0\ . \tag{12.66}$$

We denote by $\tilde{u}(r)$ the solution of Eq. (12.66) when $V(r) = 0$.

$$\left[\frac{d^2}{dr^2} + k^2 - \frac{l(l+1)}{r^2}\right]\,\tilde{u}_l(r) = 0\ . \tag{12.67}$$

We multiply from the left Eq. (12.66) by $\tilde{u}_l(r)$ and integrate from 0 to ∞. We get

$$\int_0^{\infty}\tilde{u}_l\frac{du_l^2}{dr^2}dr - \frac{2m}{\hbar^2}\int_0^{\infty}\tilde{u}_lV(r)\,u_l\,dr + \int_0^{\infty}\left(k^2 + \frac{l(l+1)}{r^2}\right)\tilde{u}_lu_l\,dr = 0\ . \tag{12.68}$$

Similarly, we multiply Eq. (12.67) from the left with u_l, integrate it from 0 to ∞, and subtract the result from Eq. (12.68). We obtain

$$\int_0^{\infty}\tilde{u}_l\,\frac{du_l^2}{dr^2}\,dr - \int_0^{\infty}u_l\,\frac{d\tilde{u}_l^2}{dr^2}\,dr = \frac{2m}{\hbar^2}\int_0^{\infty}\tilde{u}_lV(r)\,u_l\,dr\ . \tag{12.69}$$

Integrating by parts the left-hand side of this equation, we arrive at the following result

$$\left[\tilde{u}_l \frac{du_l}{dr} - u_l \frac{d\tilde{u}_l}{dr}\right]_0^\infty = \frac{2m}{\hbar^2}\int_0^\infty \tilde{u}_l V(r)\, u_l\, dr\ . \tag{12.70}$$

From the asymptotic form ($r \to \infty$) of the radial part of Eqs. (12.53) and (12.56) we can write

$$\tilde{u}_l = r j_l(kr) \to \frac{1}{k}\sin(kr - l\pi/2)\ ; \tag{12.71}$$

$$u_l = r R_l(kr) \to \frac{D_l}{k}\sin(kr - l\pi/2 + \delta_l)\ , \tag{12.72}$$

where D_l is given by Eq. (12.59). Because $R_l(r)$ is finite at $r = 0$ we must have

$$u_l(0) = \tilde{u}_l(0) = 0\ . \tag{12.73}$$

Substituting Eqs. (12.71), (12.72), and (12.73) into Eq. (12.70) gives:

$$-\frac{D_l}{k}\sin\delta_l = \frac{2m}{\hbar^2}\int_0^\infty r\, j_l(kr) V(r) u_l(r)\, dr\ , \tag{12.74}$$

where we kept the general asymptotic form for u_l in order to simplify notation. We thus obtain

$$\sin\delta_l = -k\frac{2m}{\hbar^2}\int_0^\infty r\, j_l(kr) V(r)\frac{u_l(r)}{D_l}\, dr\ . \tag{12.75}$$

Let us assume that the potential $V(r)$ is infinitesimally small. In such a case, δ_l is small, $D_l \simeq 1$, and u_l/D_l will be almost $r j_l(kr)$, the function for the free particle. With this approximation we can write Eq. (12.75) as

$$\sin\delta_l \simeq \delta_l \simeq -k\frac{2m}{\hbar^2}\int_0^\infty r^2\, j_l^2(kr) V(r)\, dr\ . \tag{12.76}$$

Since $r^2 j_l^2(kr)$ is positive, we conclude that the sign of δ_l is opposite to the sign of $V(r)$. For an attractive potential $V(r) < 0$ we get $\delta_l > 0$ and vice versa. For a repulsive potential $V(r) > 0$ we get $\delta_l < 0$. This was true for an infinitesimal potential. What about a finite one? A finite potential can be built up by successively adding up infinitesimal potentials, thus, the above argument holds true. Intuitively, incoming particles are accelerated in attractive potentials, thus they gain phase, and slowed down in a repulsive potential thus they will lose phase.

12.4.6 Low energy collisions

The usefulness of the partial wave expansion is manifest in cases where the energy of the incoming particles is relatively low, i.e., comparable to the potential energy of the target. In such cases, the quantum number l has an effective upper limit, and only the first few of the partial wave cross-sections σ_l from Eq. (12.62) will contribute significantly to the total cross-section σ. While this can be proven in a more rigorous way, here is a semiclassical model that supports the argument.

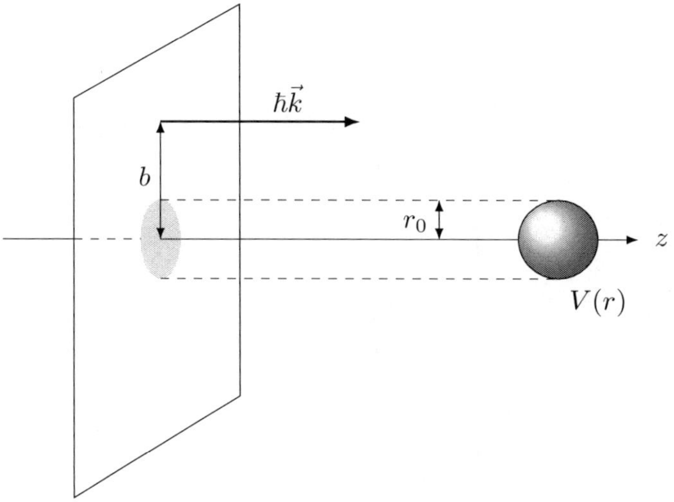

FIGURE 12.5
The semiclassical model of collision of a particle of momentum $\hbar\vec{k}$ and impact parameter b with a target of potential $V(r)$ and range r_o.

We denote by r_o the effective range of the central potential $V(r)$ and consider an incoming particle of momentum $\hbar\vec{k}$ with the impact parameter b. Its angular momentum is then

$$L = \hbar\, kb = l\hbar \ , \tag{12.77}$$

which gives a semiclassical estimate for the value of the quantum number l

$$l = kb \ . \tag{12.78}$$

where $k = \sqrt{2mE}/\hbar$. Intuitively, we do not expect much scattering when $b > r_o$. Then, the maximum quantum number l for which σ_l contributes significantly to the total σ is of the order of

$$l_{max} \simeq kr_o \ . \tag{12.79}$$

Thus, for a given r_o, the lower the energy of the incoming particle, the lower the maximum quantum number contributing to the total σ. Many cases of practical interest will be at $l = 0$, corresponding to the so-called s-wave scattering.

12.4.7 *s*-wave scattering on a finite spherical potential well

As an example, we consider the case of a finite spherical potential well shown in Fig. 12.6, also known as a "soft sphere" potential:

$$V(r) = \begin{cases} -V_0 & r < r_\text{o} \ , \\ 0 & r \geq 0 \ . \end{cases} \tag{12.80}$$

This approximates a spherically symmetric potential that is negligible outside the radius r_o, such as the screened Coulomb potential of noble gas atoms.

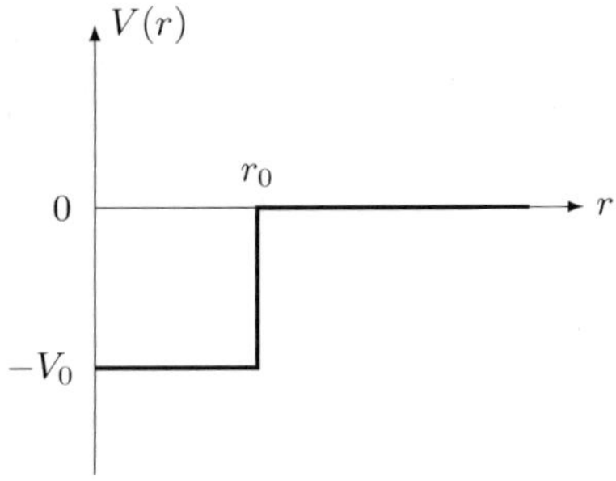

FIGURE 12.6
Soft sphere potential.

Because the potential is constant, we write the radial equation (12.45) for $r < r_\text{o}$ as following

$$\frac{d^2R_l}{dr^2} + \frac{2}{r}\frac{dR_l}{dr} + \left[\mathcal{K}^2 - \frac{l(l+1)}{r^2}\right] R_l = 0 \ , \tag{12.81}$$

where the constant target's potential was included in

$$\mathcal{K}^2 = k^2 - \frac{2m}{\hbar^2}V(r) = \frac{2m}{\hbar^2}(E + V_0) \ . \tag{12.82}$$

To ensure that wave function is regular at $r = 0$ we write the radial solution of Eq. (12.81) only in terms of the spherical Bessel functions j_l. We have

$$R_l(r) = A_l j_l(\mathcal{K}r) \ . \tag{12.83}$$

For $r > r_\text{o}$ the radial equation becomes

$$\frac{d^2R_l}{dr^2} + \frac{2}{r}\frac{dR_l}{dr} + \left[k^2 - \frac{l(l+1)}{r^2}\right] R_l = 0 \ , \tag{12.84}$$

where $k = \sqrt{2mE}/\hbar$. In this case, because both j_l and y_l are non-singular in this region, as we have already seen in Eqs. (12.55) and (12.56). The solution is a linear combination

$$R_l(r) = D_l\big[\cos\delta_l\, j_l(kr) - \sin\delta_l\, y_l(kr)\big] \ . \tag{12.85}$$

We require the usual matching of the solutions R_l and of their derivatives $R_l' \equiv dR_l/dr$ at $r = r_o$. This can be written succinctly in terms of the continuity of R_l'/R_l, sometimes referred to as the logarithmic derivative of R_l. We get

$$\Delta_l(k) \equiv \mathcal{K}\,\frac{j_l'(\mathcal{K}r_o)}{j_l(\mathcal{K}r_o)} = k\,\frac{\cos\delta_l\, j_l'(kr_o) - \sin\delta_l\, y_l'(kr_o)}{\cos\delta_l\, j_l(kr_o) - \sin\delta_l\, y_l(kr_o)} \ , \tag{12.86}$$

where the quantity $\Delta_l(k)$ depends only on k and the radial wave function for $r < r_o$. Solving Eq. (12.86) for $\tan\delta_l$, we obtain

$$\tan\delta_l = \frac{kj_l'(kr_o) - \Delta_l(k)\, j_l(kr_o)}{ky_l'(kr_o) - \Delta_l(k)\, y_l(kr_o)} \ . \tag{12.87}$$

For incoming particle with low energies, the value of k is quite small, causing the term $k^2 R_l$ in the radial Schrödinger equation to have a negligible effect on its solution. As a result, $\Delta_l(k)$ becomes essentially independent of k and can be simplified to Δ_l. As x approaches zero, the spherical Bessel functions can be expressed as

$$j_l(x) \to \frac{x^l}{(2l+1)!!} \quad ; \quad y_l(x) \to -\frac{(2l-1)!!}{x^{l+1}} \ . \tag{12.88}$$

Here, $n!!$ represents the double factorial notation that gives for an odd n, $n!! = n(n-2)\cdots 3\cdot 1$ and for an even n, $n!! = n(n-2)\cdots 4\cdot 2$.

This result is relevant for the case when $kr_o \ll 1$. Then, from Eq. (12.87), we obtain

$$\tan\delta_l \to \frac{1}{(2l+1)!!(2l-1)!!}\,\frac{l - r_o\Delta_l}{l + r_o\Delta_l + 1}\,(kr_o)^{2l+1} \ . \tag{12.89}$$

Exercise 12.7 Derive Eq. (12.89).

Eq. (12.89) shows that $\tan\delta_l$ vanishes as $(kr_o)^{2l+1}$, hence for low energies, or more precisely for $kr_o \ll 1$ cases, $\delta_l(k)$ either vanishes or becomes an integer multiple of π.

Eq. (12.86) reveals another interesting aspect of this system. Its denominator will vanish for particular values of k. Then, $\tan\delta_l \to \infty$, and consequently the phase shifts $\delta_l \to (n+1/2)\pi$, where n is an integer. When $\delta_l = \pi/2$ for example, then the partial wave cross-section σ_l reaches its maximum value for that energy:

$$\sigma_l(k) = \frac{4\pi}{k^2}\,(2l+1)\sin^2\delta_l \to \frac{4\pi}{k^2}\,(2l+1) \ . \tag{12.90}$$

Now, we go one step further. Because $kr_o \ll 1$, we ignore all $l > 0$. We are in the realm of $l = 0$ scattering, also known as the *s-wave scattering*. We consider the two cases

- $r < r_o$: In this case, from Eq. (12.51) we have $A_0 = 1$, and therefore, the asymptotic radial solution can be written as

$$R_0(\mathcal{K}r) = j_0(\mathcal{K}r) \simeq \frac{1}{\mathcal{K}r} \sin(\mathcal{K}r) . \tag{12.91}$$

- $r > r_o$: In this case from Eq. (12.59) we have $D_0 = 1$, and therefore, Eq. (12.56) gives

$$\widetilde{R}_0(kr) \simeq \frac{1}{kr} \sin(kr + \delta_0) , \tag{12.92}$$

where we denoted by $\widetilde{R}_0$ the radial function in this region.

The continuity of the wave function and its derivative yields

$$\frac{R_0'(r_o)}{R_0(r_o)} = \frac{\widetilde{R}_0'(r_o)}{\widetilde{R}_0(r_o)} . \tag{12.93}$$

After algebraic manipulation, we obtain

$$\tan(kr_o + \delta_0) = \frac{k}{\mathcal{K}} \tan(\mathcal{K}r_o) , \tag{12.94}$$

or

$$\delta_0 = \arctan\left[\frac{k}{\mathcal{K}} \tan(\mathcal{K}r_o)\right] - kr_0 , \tag{12.95}$$

which is an exact expression for the s-wave δ_0 phase shift.

We conclude this section with one more approximation. If the energy of the incoming particles is close to zero, then we also have

$$\frac{k}{\mathcal{K}} = \sqrt{\frac{E}{E + V_0}} \to 0 , \tag{12.96}$$

and consequently, for values of $\mathcal{K}r_o$ that render relatively small values of $\tan(\mathcal{K}r_o)$, we can write

$$\arctan\left[\frac{k}{\mathcal{K}} \tan(\mathcal{K}r_o)\right] \simeq \frac{k}{\mathcal{K}} \tan(\mathcal{K}r_o) . \tag{12.97}$$

Substituting Eq. (12.97) into (12.95), and keeping in mind that we are in the regime where $kr_o \ll 1$, we obtain

$$\delta_0 \simeq kr_0 \left[\frac{\tan(\mathcal{K}r_o)}{\mathcal{K}r_o} - 1\right] \simeq \sin\delta_0 . \tag{12.98}$$

Using Eq. (12.12) where $f(\theta)$ is given by Eq. (12.61) with only one value for l, namely $l = 0$, we obtain for the differential cross-section

$$\frac{d\sigma}{d\Omega} \simeq \frac{1}{k^2} \sin^2 \delta_0 \ . \tag{12.99}$$

Since $d\sigma/d\Omega$ is isotropic, the total cross-section becomes

$$\sigma \simeq \frac{4\pi}{k^2} \sin^2 \delta_0 \simeq 4\pi r_0^2 \left[\frac{\tan(\mathcal{K} r_\text{o})}{\mathcal{K} r_\text{o}} - 1 \right] \ . \tag{12.100}$$

This is an interesting result that the reader first met when calculating the transmission coefficient for the one-dimensional finite square well. Similarly to the one-dimensional case, in this case, $\sigma \to 0$ when $\tan(\mathcal{K} r_\text{o}) = \mathcal{K} r_\text{o}$, or when $\delta_0 = \pi, 2\pi, \ldots$. This is a remarkable effect which was first observed experimentally as a dip in the scattering probability of low-energy electrons traversing a container with a noble gas. The dip occurred when the electrons had energies of particular values. Today, the effect is known as the Ramsauer-Townsend effect, after Carl Ramsauer and John Sealy Townsend, who, in 1921, independently studied the collisions between low-energy electrons and noble gas atoms, such as argon, krypton, or xenon. For example, for electrons moving in xenon[3], the first dip in the collision probability was observed for energies near $1\,eV$. With the advent of quantum mechanics, this puzzling effect was finally understood.

12.5 Problems

12.1 In a scattering experiment the detector is located far away from the target where $V = 0$, and thus the Schrödinger equation for the reduced radial function $u = rR$ becomes

$$-\frac{\hbar^2}{2m}\frac{d^2 u}{dr^2} + \frac{\hbar^2 \ell(\ell+1)}{2mr^2} u = Eu \ .$$

Solve this equation in the large r limit and show that the radial function $R(r)$ can be asymptotically written as

$$R(r) = \frac{e^{ikr}}{r} \quad ; \quad k = \sqrt{\frac{2mE}{\hbar^2}} \ .$$

12.2 Derive Eq. (12.6) using the Schrödinger equation and its complex conjugate.

[3] The experimental results are summarized by R. Kollath in Phys. Zeits. **31**, (1931) 985.

12.3 Using Gauss's theorem for a finite volume show that Eq. (12.5) represents the change of the probability density due to a probability current $\vec{j}$ flowing through the surface enclosing that volume.

12.4 Using Born approximation, calculate the scattering amplitude $f(q)$ and the total cross section σ for a soft sphere described by the potential

$$V(r) = \begin{cases} V_0 & r \leq r_\text{o} \,, \\ 0 & r > r_\text{o} \,. \end{cases}$$

12.5 Using Born approximation, calculate the differential cross-section for the scattering of a particle of mass m and charge e that collides perpendicularly with an electric dipole consisting of two charges e and $-e$ placed at a distance d. Find the angle θ for which $d\sigma/d\Omega$ is maximal. Using a software of your choice, plot $d\sigma/d\Omega$ vs. θ.

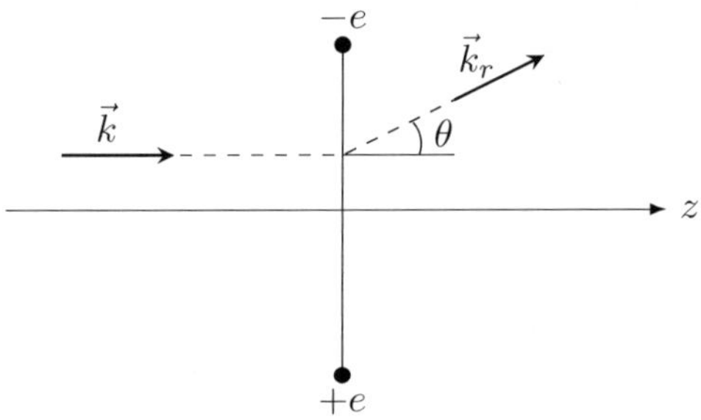

FIGURE 12.7
Figure for Problem 12.5: charged particle scattering on an electric dipole.

12.6 Derive Eq. (12.94). By explicit calculation show that the same result is obtained by using the reduced radial wave function $u(r) = rR(r)$ and imposing the continuity of its logarithmic derivative at $r = r_\text{o}$.

12.7 For the case of low energies, i.e., where $kr_\text{o} \ll 1$, calculate the phase shift δ_0 and the total cross section σ for a repulsive soft sphere described by the potential

$$V(r) = \begin{cases} V_0 & r \leq r_\text{o} \,, \\ 0 & r > r_\text{o} \,. \end{cases}$$

where $V_0 > 0$. Compare your results with the case of the attractive sphere, $V_0 < 0$.

13

Superposition, Entanglement, and Quantum Information

13.1 Introduction

In recent years, due to advances in technology that allow for precise manipulation of quantum states, we can control superposition and entanglement of quantum states for an increasing number of states. These advances sparked an explosive growth of new technologies that led to an effective implementation of quantum encryption protocols and the development of quantum computers able to process over a thousand qubits. This rapid development, sometimes referred to as "the second quantum revolution," was made possible by the development of precise laser manipulation, increased precision in solid state experiments, and the promise of novel computational tools that could solve problems currently beyond the reach of supercomputers. We will start this chapter by introducing the fundamental unit of quantum information, the quantum bit or *qubit*, for short. We will then discuss entanglement and the conceptual conundrum it raised, the EPR paradox, and Bell's theorem. Afterward, we delve into modern applications of these fundamental ideas, such as quantum teleportation, quantum cryptography, and elements of quantum computation.

13.2 Qubits

We will consider two-state systems, such as an electron that can have spin-up or spin-down with respect to an external magnetic field, or two states of polarization of a photon, which for example, can be circularly left or right polarized. We will not need to know what type of system we have, only that it is composed of one or more identical particles, each having two possible states which we will conveniently denote by $|0\rangle$ and $|1\rangle$.

A *qubit* is a linear superposition of $|0\rangle$ and $|1\rangle$:

$$|q\rangle = \alpha\,|0\rangle + \beta\,|1\rangle \ , \tag{13.1}$$

DOI: 10.1201/9781003174967-13

where the complex coefficients α and β satisfy $|\alpha|^2+|\beta|^2 = 1$. Unlike a classical bit which can take only two values, 0 and 1, a qubit can be in an infinite number of states. However, upon making a measurement, a qubit collapses into one of the two possible states: $|0\rangle$ or $|1\rangle$. We do not know *a priori* in which final state the system ends up, but we know the probabilities of each: $|\alpha|^2$ for state $|0\rangle$ and $|\beta|^2$ for $|1\rangle$.

How can we visualize a qubit? Because α and β are complex, there are four real numbers that determine a qubit. Due to the normalization condition $|\alpha|^2 + |\beta|^2 = 1$ we need only three, γ, δ and θ, which we can write as

$$\alpha = e^{i\gamma} \cos\frac{\theta}{2} \quad ; \quad \beta = e^{i\delta} \sin\frac{\theta}{2} \ . \tag{13.2}$$

Because an overall phase is irrelevant, we further denote $\varphi = \delta - \gamma$, and obtain

$$|q\rangle = \cos\frac{\theta}{2}\,|0\rangle + e^{i\varphi} \sin\frac{\theta}{2}\,|1\rangle \ , \tag{13.3}$$

where $0 \leq \theta \leq \pi$ and $0 \leq \varphi \leq 2\pi$. If we identify θ and ϕ with the spherical coordinates' angles, this parametrization yields a one-to-one correspondence between the qubit $|q\rangle$ and the unit vector

$$\vec{n} = \vec{\imath} \sin\theta\cos\varphi + \vec{\jmath} \sin\theta\sin\varphi + \vec{k}\cos\theta \ , \tag{13.4}$$

denoting a point on the surface of the sphere of radius 1, known as the *Bloch sphere.*

In this representation, the state $|0\rangle$ is at the north pole, and $|1\rangle$ at the south pole of the Bloch sphere. Two other states, labeled $|+\rangle$ and $|-\rangle$ will be of importance later. They are defined as

$$|+\rangle = \frac{1}{\sqrt{2}}\,|0\rangle + \frac{1}{\sqrt{2}}\,|1\rangle \quad ; \quad |-\rangle = \frac{1}{\sqrt{2}}\,|0\rangle - \frac{1}{\sqrt{2}}\,|1\rangle \ . \tag{13.5}$$

They correspond to $\theta = \pi/2$ and $\varphi = 0$ for $|+\rangle$, and $\theta = \pi/2$ and $\varphi = \pi$ for $|-\rangle$, as illustrated in Fig. 13.1. In general, two orthogonal qubits are represented by antipodal points on the Bloch sphere.

Exercise 13.1 Using Eq. (13.3), verify that two qubits corresponding to two antipodal points on the Bloch sphere are orthogonal.

Qubits live in a two-dimensional Hilbert space, and if we represent them in matrix form as

$$|0\rangle \to \begin{pmatrix} 1 \\ 0 \end{pmatrix} \quad ; \quad |1\rangle \to \begin{pmatrix} 0 \\ 1 \end{pmatrix} ,$$

then

$$|q\rangle \to \begin{pmatrix} \alpha \\ \beta \end{pmatrix} .$$

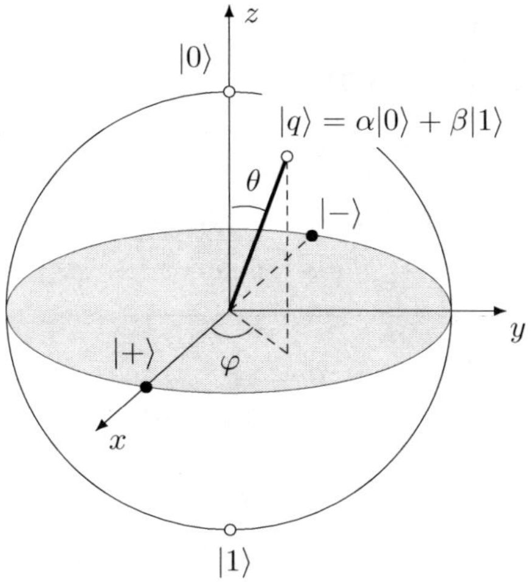

FIGURE 13.1
The representation of a qubit $|q\rangle = \alpha\,|0\rangle + \beta\,|1\rangle$ on the Bloch sphere. The antipodal points correspond to orthogonal qubits.

If we have a two-particle system, then we work with two qubits, $|q_1\rangle$ and $|q_0\rangle$. Mathematically, we represent such a system by the tensor product $|q_1\rangle \otimes |q_0\rangle$. To simplify notation, we omit the tensor product symbol and write it as $|q_1\rangle\,|q_0\rangle$ or $|q_1 q_0\rangle$. Note that by considering two qubits, the dimensionality of the Hilbert space has increased to four:

$$|q_1 q_0\rangle \rightarrow \begin{pmatrix} \alpha_1 \\ \beta_1 \end{pmatrix} \otimes \begin{pmatrix} \alpha_0 \\ \beta_0 \end{pmatrix} = \begin{pmatrix} \alpha_1 \begin{pmatrix} \alpha_0 \\ \beta_0 \end{pmatrix} \\ \beta_1 \begin{pmatrix} \alpha_0 \\ \beta_0 \end{pmatrix} \end{pmatrix} = \begin{pmatrix} \alpha_1\alpha_0 \\ \alpha_1\beta_0 \\ \beta_1\alpha_0 \\ \beta_1\beta_0 \end{pmatrix} .$$

For example,

$$|00\rangle \rightarrow \begin{pmatrix} 1 \\ 0 \\ 0 \\ 0 \end{pmatrix} .$$

The tensor product can be generalized to three or more qubits. As a result, the dimension of the corresponding space increases exponentially with the number of qubits. For n qubits, we have a 2^n-dimensional vector space.

As we will see later, computational needs often require more than one qubit. In the following, we will use the ordering convention in which the n-th qubit is on the left side of the tensor product:

$$|q_n\rangle \otimes |q_{n-1}\rangle \otimes \cdots \otimes |q_0\rangle \equiv |q_n q_{n-1} \cdots q_0\rangle \ . \tag{13.6}$$

13.3 Entanglement, EPR, and Bell's theorem

The superposition of quantum states was the first signal that, at microscopic scales, nature behaves vastly differently from its "intuitive" macroscopic behavior. Entanglement added to that sense of wonder and has ignited vigorous scientific debates that paved the way for today's quantum technologies.

13.3.1 Entanglement

As we have seen, a qubit is a superposition of two orthogonal states $|q\rangle = \alpha|0\rangle + \beta|1\rangle$. Now, we consider two qubits. They can be in particular states such as $|00\rangle$ or $|11\rangle$, or in general they can be a superposition of all four states

$$|\psi\rangle = \alpha|00\rangle + \beta|01\rangle + \gamma|10\rangle + \delta|11\rangle \ , \tag{13.7}$$

where, normalization requires $|\alpha|^2 + |\beta|^2 + |\gamma|^2 + |\delta|^2 = 1$. A multi-particle quantum state is called *entangled* if it cannot be written as a product of states. For example, the state of Eq. (13.7) is entangled.

Four particular entangled states, known as *Bell states*, [36, 49] are widely used in quantum information. They are

$$\begin{aligned} |\Phi^+\rangle &= \frac{1}{\sqrt{2}}|00\rangle + \frac{1}{\sqrt{2}}|11\rangle \quad , \quad |\Phi^-\rangle = \frac{1}{\sqrt{2}}|00\rangle - \frac{1}{\sqrt{2}}|11\rangle \\ |\Psi^+\rangle &= \frac{1}{\sqrt{2}}|01\rangle + \frac{1}{\sqrt{2}}|10\rangle \quad , \quad |\Psi^-\rangle = \frac{1}{\sqrt{2}}|01\rangle - \frac{1}{\sqrt{2}}|10\rangle \ . \end{aligned} \tag{13.8}$$

Note that none of these can be decomposed as a tensor product of simpler states.

13.3.2 Einstein-Podolski-Rosen paradox

Entangled states exhibit apparently contradictory behaviors. Einstein was one of the prominent physicists who believed that the probabilistic nature of quantum mechanics was due to our ignorance of some unknown degrees of freedom or "hidden variables." While acknowledging the success of the quantum mechanics, Einstein thought that in its current form, quantum mechanics was incomplete.

To prove this incompleteness, Einstein, Podolski, and Rosen (EPR), in a 1935 paper [12], explored an essential consequence of entanglement. In short, they concluded that if entanglement is possible, then causality must be broken because entanglement, as described by the orthodox quantum mechanics, implies instantaneous action at the distance between the two particles. We are going to follow the same line of reasoning, using a simpler setup initially proposed by David Bohm [7, 9]. Consider two spin-1/2 entangled particles A

and B in the Bell state $|\Psi^-\rangle$:

$$\begin{aligned} |\Psi^-\rangle &= \frac{1}{\sqrt{2}}|01\rangle - \frac{1}{\sqrt{2}}|10\rangle \ , \\ &= \frac{1}{\sqrt{2}}\left|\tfrac{1}{2}\right\rangle_A\left|-\tfrac{1}{2}\right\rangle_B - \frac{1}{\sqrt{2}}\left|-\tfrac{1}{2}\right\rangle_A\left|\tfrac{1}{2}\right\rangle_B \ , \end{aligned} \tag{13.9}$$

where we explicitly indicated that $|0\rangle$ is the spin-up state $|\frac{1}{2}\rangle$ and $|1\rangle$ the spin-down $|-\frac{1}{2}\rangle$. We will make contact with the experiment by assuming that A and B are the electron and respectively the positron generated by the decay of a neutral pion π^0 at rest:

$$\pi^0 \to e^- + e^+ \ .$$

Because the electron and the positron have the same mass, the conservation of momentum requires that they fly away in opposite directions with the same speed, as measured in the lab's frame of reference, where the pion was at rest. They also must be in a singlet spin state to preserve the overall zero spin of the system.

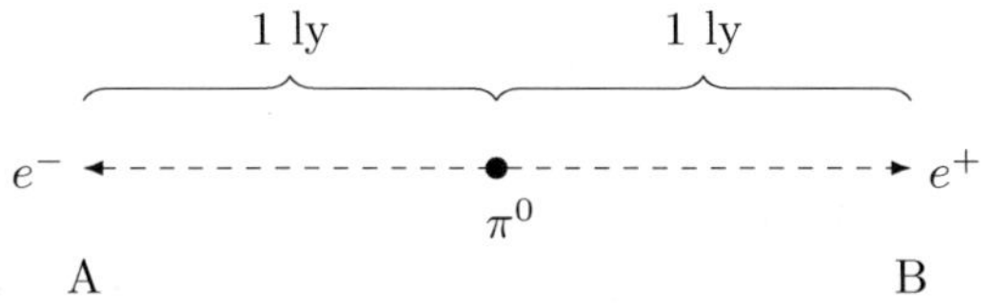

FIGURE 13.2
Entangled pair of electron-positron resulting from the decay of a neutral meson π^0 at rest in the lab frame.

Suppose we measure the spin of the electron, which could be arbitrarily far away at point A. According to quantum mechanics, we don't know *a priori* the result of the measurement. There is a 50% chance of obtaining either $|\frac{1}{2}\rangle_A$ or $|-\frac{1}{2}\rangle_A$. However, if, for example, we obtained $|\frac{1}{2}\rangle_A$, then the positron at B can only be in the state $|-\frac{1}{2}\rangle_B$. Similarly, if we measured $|-\frac{1}{2}\rangle_A$, then the positron must be in $|\frac{1}{2}\rangle_B$. The perplexing fact is that such a result is true, regardless of the distance between the two entangled particles. For illustration purposes, Fig. 13.2 shows the two particles separated by a distance of 2 light-years.[1]

According to EPR, such a result is unacceptable because it would imply some sort of supraluminal interaction between the two particles. The EPR authors thought that in reality the electron and the positron were in a definite state from the beginning: either $|\frac{1}{2}\rangle_A|-\frac{1}{2}\rangle_B$ or $|-\frac{1}{2}\rangle_A|\frac{1}{2}\rangle_B$. The reason we wouldn't know this initial state was due to the incompleteness of quantum

[1]Of course, in ideal conditions, the distance between the two particles plays no role.

mechanics. A complete theory, they believed, must include some additional, *hidden variables.* These variables would precisely give the system's initial state, thus removing the need for probability in quantum mechanics. In such a scenario, the problem would be solved because if the electron was in $|\frac{1}{2}\rangle_A$, for example, then the positron must have been in $|-\frac{1}{2}\rangle_B$, and no faster than light action would ever be needed.

We would like to emphasize again, that the orthodox point of view of quantum mechanics is rather pragmatic; it assumes that nature is inherently probabilistic, and thus, no hidden variables are needed. The two particles are in a superposition of states: $|\frac{1}{2}\rangle_A |-\frac{1}{2}\rangle_B$ and $|-\frac{1}{2}\rangle_A |\frac{1}{2}\rangle_B$. When we measure the spin of one particle, for example the electron at A, we interact with the entire two-particle state $|\Psi^-\rangle$, which collapses into either $|\frac{1}{2}\rangle_A |-\frac{1}{2}\rangle_B$ or $|-\frac{1}{2}\rangle_A |\frac{1}{2}\rangle_B$.

Short of an experimental pathway able to falsify one of the two competing scenarios, physicists continued to develop in parallel to the orthodox approach, quantum theories with hidden variables [7, 8], that were able to explain the same results as the orthodox quantum mechanics, while preserving locality and causality of the theory. This changed after John S. Bell demonstrated that specific predictions of quantum mechanics are incompatible with local hidden variable theories, thus challenging the plausibility of a deterministic underpinning of quantum mechanics and reinforcing the probabilistic framework of orthodox quantum mechanics.

13.3.3 Bell's theorem

In a 1964 paper, Bell [4] proposed an experimental procedure that provided the falsifying criterion needed to discern between theories with hidden variables and the orthodox quantum mechanics. This result is known today as *Bell's theorem* and it provided a pathway to rule out all hidden variable theories, regardless of their inner workings. Experiments inspired by Bell's theorem proved that any local hidden variable theory is incompatible with quantum mechanics. The universe is indeed probabilistic at its core, and there are no hidden variables.

Bell's proof of the main result used the EPR-Bohm setting. A source of entangled particles (here we consider it generates electron-positron pairs)[2] is placed in the middle of two detectors located at A and B, as illustrated in Fig. 13.3.

The identical detectors measure the spin component along the unit vectors $\vec{a}$ and $\vec{b}$ respectively. Detector A gives ± 1 if the spin of the incoming electron is $\pm\hbar/2$ and similarly, the detector B gives ± 1 if the spin of the incoming positron is $\pm\hbar/2$. Next, we fix the detectors' orientations $\vec{a}$ and $\vec{b}$ and calculate $\eta(\vec{a}, \vec{b})$ which is the average of the products of detectors' readings for a large number of entangled pairs. Because the two particles are in a singlet spin state (zero

[2]The nature of the entangled particles is irrelevant. The first successful experiments were done with entangled photons generated via non-linear optical techniques.

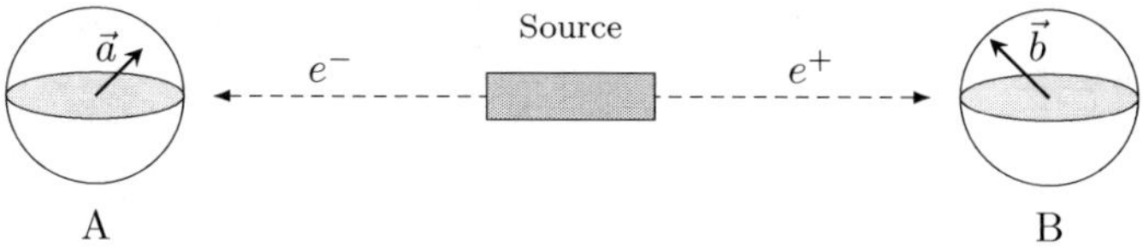

FIGURE 13.3
Bell's theorem. The detectors are arbitrarily oriented along $\vec{a}$ and $\vec{b}$.

total spin) when the detectors are parallel, $\vec{a} = \vec{b}$, we have

$$\eta(\vec{a}, \vec{a}) = -1 \ , \tag{13.10}$$

and respectively, when they are anti-parallel, $\vec{a} = -\vec{b}$

$$\eta(\vec{a}, -\vec{a}) = 1 \ . \tag{13.11}$$

In general,

$$-1 \leq \eta(\vec{a}, \vec{b}) \leq 1 \ . \tag{13.12}$$

The spin operator for one-half spin particles is $\widehat{\vec{S}} = \frac{\hbar}{2}\vec{\sigma}$, where $\vec{\sigma}$ are Pauli matrices. Then, the orthodox quantum mechanics predicts:

$$\eta(\vec{a}, \vec{b}) = \langle (\vec{\sigma}^A \cdot \vec{a})\, (\vec{\sigma}^B \cdot \vec{b}) \rangle = -\vec{a} \cdot \vec{b} \ , \tag{13.13}$$

where, upper indices A and B refer to the electron and the positron, respectively. Before going any further, we stop to analyze this equation. Detector A measures the spin component $\hat{S}_a^A = \frac{\hbar}{2}\vec{\sigma}^A \cdot \vec{a}$ along the $\vec{a}$ unit vector. It yields ± 1 if the spin component is $\pm\hbar/2$. Thus, the operator $\vec{\sigma}^A \cdot \vec{a}$ models [51] the output of detector A. Similarly, for detector B. The second part of part of Eq. (13.13) is left as an exercise.

Exercise 13.2 Derive Eq. (13.13) knowing that the system is in the singlet spin state given by Eq. (13.9).

Now, we will analyze the same experimental setup from the point of view of a general hidden-variable theory. Bohm and Aharonov [9] argued that the singlet quantum state of Eq. (13.9) is the result of not taking into account additional hidden variables, that, once determined by a complete theory, would lift the randomness from the quantum measuring process. The final result should be the same as in Eq. (13.13). But, as Bell proved, this is not possible.

In the proof below we will closely follow Bell's argument [4]. We assume that the additional parameters λ complete the quantum theory. The exact nature of λ is irrelevant. The parameter could be a single one which is continuous or discrete, or it could be a family of parameters, etc. For conciseness, we will write as if it were a single continuous parameter, with an unknown probability distribution function $\rho(\lambda)$. This is a non-negative function obeying the normalization condition:

$$\int \rho(\lambda)\, d\lambda = 1 \ ,$$

where the integral covers the entire parameter space. Although different hidden-variable theories could have different probability distributions, this proof doesn't need anything specific, thus its conclusions will hold true for all such theories.

According to [9], the result given by the detector A, which measures $\vec{\sigma}^A \cdot \vec{a}$, is determined only by $\vec{a}$ and λ. We then write:

$$A(\vec{a}, \lambda) = \pm 1 \ . \tag{13.14}$$

Similarly, the output of B depends only on $\vec{b}$ and λ

$$B(\vec{b}, \lambda) = \pm 1 \ . \tag{13.15}$$

One essential tenet of the argument in Ref. [9] is that the result measured by B does not depend on the orientation of $\vec{a}$, and reciprocally, the result of A does not depend on the orientation of $\vec{b}$.

The expectation value of the product of the readings of the two detectors is then

$$P(\vec{a}, \vec{b}) = \int A(\vec{a}, \lambda) B(\vec{b}, \lambda) \rho(\lambda)\, d\lambda \ , \tag{13.16}$$

which, if different from $\eta(\vec{a}, \vec{b})$ of Eq. (13.13), would provide the falsifying criterion that would make the distinction between hidden-variable theories and the orthodox quantum mechanics. When the detectors are aligned, the overall zero spin requires

$$A(\vec{a}, \lambda) = -B(\vec{a}, \lambda) \ , \tag{13.17}$$

and thus, Eq. (13.16) becomes

$$P(\vec{a}, \vec{b}) = -\int A(\vec{a}, \lambda) A(\vec{b}, \lambda) \rho(\lambda)\, d\lambda \ . \tag{13.18}$$

Considering a third, arbitrary, unit vector $\vec{c}$, we have

$$\begin{aligned} P(\vec{a}, \vec{b}) \ - \ P(\vec{a}, \vec{c}) &= -\int \Big[A(\vec{a}, \lambda) A(\vec{b}, \lambda) - A(\vec{a}, \lambda) A(\vec{c}, \lambda)\Big] \rho(\lambda)\, d\lambda \\ &= -\int A(\vec{a}, \lambda) A(\vec{b}, \lambda) \Big[1 - A(\vec{b}, \lambda) A(\vec{c}, \lambda)\Big] \rho(\lambda)\, d\lambda \ , \end{aligned} \tag{13.19}$$

where in the last line we used the fact that $A(\vec{b}, \lambda) A(\vec{b}, \lambda) = 1$. Since $|A(\vec{a}, \lambda) A(\vec{b}, \lambda)| = 1$, we have

$$\begin{aligned} -\int \Big[1 - A(\vec{b}, \lambda) A(\vec{c}, \lambda)\Big] \rho(\lambda) d\lambda \le P(\vec{a}, \vec{b}) - P(\vec{a}, \vec{c}) \\ \le \int \Big[1 - A(\vec{b}, \lambda) A(\vec{c}, \lambda)\Big] \rho(\lambda) d\lambda \end{aligned}$$

or, because $\Big[1 - A(\vec{b},\lambda)A(\vec{c},\lambda)\Big]\rho(\lambda) \geq 0$, we can rewrite everything more concisely as

$$|P(\vec{a},\vec{b}) - P(\vec{a},\vec{c})| \leq \int \Big[1 - A(\vec{b},\lambda)A(\vec{c},\lambda)\Big]\rho(\lambda)\,d\lambda\ . \tag{13.20}$$

Next, we use the normalization condition for $\rho(\lambda)$, in the left-hand side of Eq. (13.20). We obtain

$$L.h.s. = 1 - \int A(\vec{b},\lambda)A(\vec{c},\lambda)\rho(\lambda)\,d\lambda = 1 + P(\vec{b},\vec{c})\ . \tag{13.21}$$

We arrived to the celebrated *Bell's inequality*

$$1 + P(\vec{b},\vec{c}) \geq |P(\vec{a},\vec{b}) - P(\vec{a},\vec{c})|\ . \tag{13.22}$$

At this point we see that the orthodox quantum mechanics is not compatible with Bell's inequalities. Replacing $P(\vec{a},\vec{b})$ with $\eta(\vec{a},\vec{b}) = -\vec{a}\cdot\vec{b}$, quantum mechanics would give for any unit vectors $\vec{a}$, $\vec{b}$, and $\vec{c}$:

$$1 - \vec{b}\cdot\vec{c} \geq |\vec{a}\cdot\vec{c} - \vec{a}\cdot\vec{b}|\ . \tag{13.23}$$

For example, we consider the orientations of the detectors such that $\vec{a}\cdot\vec{c} = 0$ and $\vec{a}\cdot\vec{b} = \vec{b}\cdot\vec{c} = 1/\sqrt{2}$, as illustrated in Fig. 13.4.

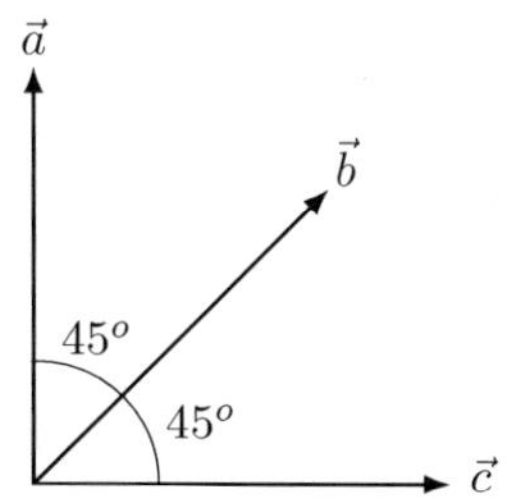

FIGURE 13.4
Violation of Bell's inequality for a particular orientation of the detectors.

Then, Eq. (13.23) gives a clear violation of the Bell's inequality:

$$0.293 \not\geq |0 - 0.707| - 0.707\ .$$

Soon after Bell's paper, experiments were performed in different experimental setups. One of the most definitive experiments at the time was performed by Aspect, Grangier, and Roger, in 1982 [2]. It showed that Bell's inequality was violated by a large margin. Thus, quantum predictions violate the constraints required by the local realism, such as "common sense" assumptions that the measurement at A is independent of the measurement at B.

As technology evolved and more sensitive sensors became available, the eventual loopholes related to detector detection time, spurious interaction between detectors, between detectors and source, etc., were gradually eliminated. Today, Bell's inequality can be verified in an undergraduate laboratory. [3]

13.4 Pure states, mixed states, and the density operator

13.4.1 Pure states

Up to now, we have looked at phenomena with a maximum amount of information. What does it mean? Consider for example a beam of electrons in state $|\vec{n}, +\rangle$ that exits an ideal Stern-Gerlach apparatus all having their spin parallel with the detector's magnetic field, which is oriented along the unit vector $\vec{n}$. We have maximal information, in the sense that we know that all these electrons are in spin-up states with respect to the detector's field. States of maximal information are called *pure states*. For pure states, we can always design an experiment [13] that gives results that are 100 % predictable. For the example of the outgoing electrons considered above, a subsequent Stern-Gerlach detector with its magnetic field parallel to the first one will always produce a single output, reinforcing our knowledge that all electrons are in spin-up states with respect to that field.

Repeating the same experiment, will certainly produce the same result, because the system was left undisturbed. This is a filter type experiment, and it can be used to prepare a system in a pure state. In the example above, adding subsequent Stern-Gerlach apparatuses will not change the state $|\vec{n}, +\rangle$ of the electrons in the beam. Such an experiment characterizes uniquely the state of the system, and it is called a *complete experiment*. Note that "complete" here has a limited meaning because it may describe only a partial set of variables describing the system. In our example, the $(SG)_{\vec{n}}$ apparatus gives no information about the momentum or the energy of the outgoing electrons.

13.4.2 Mixed states

There are quantum mechanical systems for which there is no complete experiment to give unique results with certainty. For example, in a Stern-Gerlach experiment carried out in a laboratory, the silver atoms ejected by the oven will have a thermal energy distribution, and consequently, the atoms in the input beam will have a wide range of energies. Additionally, the magnetic field may not be as sharply defined in a direction $\vec{n}$ as assumed in the ideal case. Thus, not all silver atoms in the two emerging beams will have the total spin uniquely oriented as parallel or anti-parallel with $\vec{n}$. The beam corresponding to the ideal $|\vec{n}, +\rangle$ channel may have $|\vec{n}, -\rangle$ oriented atoms, and vice versa. A subsequent $(SG)_{\vec{n}}$ apparatus as depicted in Fig. 13.5 may have some silver atoms exiting the $|\vec{n}, -\rangle$ channel as well. In such cases, we say that the information is less than maximal. By this, we mean that no complete experiment will give a uniquely determined outcome. Nevertheless, we can gather data statistically and thus characterize all possible outcomes by giving the probabilities p_i of their occurrence.

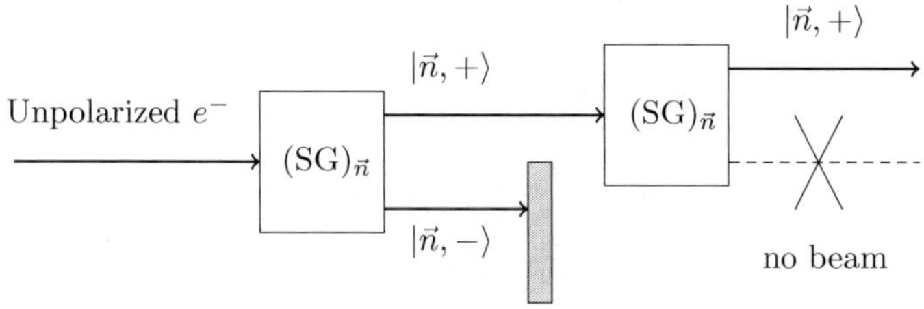

FIGURE 13.5
A pure state $|\vec{n},+\rangle$ created by an ideal Stern-Gerlach apparatus $(SG)_{\vec{n}}$ with the magnetic field oriented along direction $\vec{n}$. The filtered electrons have spins up (parallel) with respect to $\vec{n}$. A second $(SG)_{\vec{n}}$ will measure that all incoming electrons also have spin up.

States which are not pure are called *mixed states* . They are described by an incoherent superposition of pure states, which is determined as follows. First, we experimentally measure the probability p_i for of each pure state by taking repeated measurements and finding the frequency of their occurrence. Then, we assign to each of the pure states a "weight" which is equal to their probability.

Finally, we must emphasize that there is no unique description of a mixed state as an incoherent superposition of pure states. For example, unpolarized light can be described as a mixture of two linear polarizations or two circular ones.

From a practical point of view, we describe a system in terms of either its method of preparation or its history [13]. Details regarding the state's preparation can be substituted with adequate experimental observations conducted on a set of identically prepared systems. In cases where we completely lack information concerning a particular variable, we express this by specifying that certain operators have zero expectation values. For example, if the orientation of a particle's spin is unknown, the expectation value for each spin component is zero.

13.4.3 The density operator

The description of quantum states using state vectors is well-suited for pure states. There is another, equivalent, way of describing these states using the formalism of density operators [10]. One advantage of the density operators approach is that it can be readily generalized to mixed states.

13.4.3.1 The density operator for pure states

We begin by considering a pure state $|\psi(t)\rangle$ expressed in terms of an orthonormal basis $\{|n\rangle\}$

$$|\psi(t)\rangle = \sum_n c_n(t)\,|n\rangle \ , \tag{13.24}$$

where the complex coefficients $c_n(t) = \langle n|\psi(t)\rangle$ satisfy the normalization condition

$$\sum_n |c_n(t)|^2 = 1 \; . \tag{13.25}$$

The expectation value of an observable $\hat{A}$ is

$$\langle A\rangle(t) = \langle\psi(t)|\hat{A}|\psi(t)\rangle = \sum_{k,n} \langle\psi(t)|k\rangle \langle k|\,\hat{A}\,|n\rangle \langle n|\psi(t)\rangle = \sum_{k,n} c_k^*(t)c_n(t)A_{kn} \; ,$$

where $A_{kn} = \langle k|\hat{A}|n\rangle$ are the matrix elements of $\hat{A}$ in basis $\{|n\rangle\}$. Note that

$$\langle n|\psi(t)\rangle \langle\psi(t)|k\rangle = c_k^*(t)c_n(t) \; , \tag{13.26}$$

can be viewed as the matrix element of a a new operator $\hat{\rho}$, called the *density operator*[3]

$$\hat{\rho}(t) = |\psi(t)\rangle \langle\psi(t)| \; . \tag{13.27}$$

With this definition, the density operator is represented by the *density matrix* $\rho(t)$ whose elements are

$$\rho_{nk}(t) = \langle n|\psi(t)\rangle \langle\psi(t)|k\rangle = c_k^*(t)c_n(t) \; . \tag{13.28}$$

Knowing the density operator allows us to completely characterize the state of the system, i.e., it allows us to obtain the same results that can be calculated from $|\psi(t)\rangle$. We begin with the normalization condition Eq. (13.25), which written in terms of the density matrix ρ, indicates that the sum of its diagonal elements, also known as trace, is equal to 1, i.e.,

$$\sum_n |c_n(t)|^2 = \sum_n \rho_{nn}(t) = \mathrm{Tr}\,\hat{\rho}(t) = 1 \; . \tag{13.29}$$

Next, we calculate the expectation value of $\hat{A}$, and obtain

$$\begin{aligned} \langle A\rangle(t) &= \sum_{k,n} \rho_{nk}(t)A_{kn} \\ &= \sum_{k,n} \langle n|\hat{\rho}(t)|k\rangle \langle k|\hat{A}|n\rangle = \sum_n \langle n|\hat{\rho}(t)\hat{A}|n\rangle \\ &= \mathrm{Tr}\,\{\hat{\rho}(t)\hat{A}\} \; . \end{aligned} \tag{13.30}$$

Then, we use Schrödinger equation to calculate the time evolution of the density operator. We get

$$\begin{aligned} \frac{d}{dt}\hat{\rho}(t) &= \left(\frac{d}{dt}\,|\psi(t)\rangle\right) \langle\psi(t)| + |\psi(t)\rangle \left(\frac{d}{dt}\,\langle\psi(t)|\right) \\ &= \frac{1}{i\hbar}\hat{H}(t)\,|\psi(t)\rangle \langle\psi(t)| + \frac{1}{(-i\hbar)}\,|\psi(t)\rangle \langle\psi(t)|\,\hat{H}(t) \\ &= \frac{1}{i\hbar}[\hat{H}(t), \hat{\rho}(t)] \; . \end{aligned}$$

[3]Note that $\hat{\rho}(t)$ is not a *bona fide* Schrödinger operator since it is time-dependent, whereas all operators in the Schrödinger picture are time independent.

To simplify notation, henceforth we will not explicitly write the time dependence, reinstating it only where it is needed for clarity. Using Eq. (13.27) we find that the density operator ρ is Hermitian:

$$\hat{\rho}^\dagger = (|\psi\rangle\langle\psi|)^\dagger = |\psi\rangle\langle\psi| = \hat{\rho} \; . \tag{13.31}$$

In addition, we have

$$\hat{\rho}^2 = \hat{\rho}\hat{\rho} = |\psi\rangle\langle\psi|\psi\rangle\langle\psi| = |\psi\rangle\langle\psi| = \hat{\rho} \; ,$$

or,

$$\hat{\rho}^2 = \hat{\rho} \; . \tag{13.32}$$

This last property is called *idempotence.* As we will see shortly, the density operator is idempotent only for pure states.

To summarize, we found the following properties for the density operator for pure states:

1) $\mathrm{Tr}\,\hat{\rho} = 1$ Conservation of probability

2) $\langle A\rangle = \mathrm{Tr}\,\{\hat{\rho}\hat{A}\}$ The expectation value of observable $\hat{A}$

3) $\dfrac{d}{dt}\hat{\rho} = \dfrac{1}{i\hbar}[\hat{H}, \hat{\rho}]$ The time evolution of $\hat{\rho}$

4) $\hat{\rho}^\dagger = \hat{\rho}$ Hermiticity

5) $\hat{\rho}^2 = \hat{\rho}$ Idempotence

Example 13.1 The density matrix for spin-1/2 particles
Consider the case of a spin-1/2 particle, and denote by $\{|z,+\rangle, |z,-\rangle\}$ the eigenvectors of the spin operator $\hat{S}_z$. Then, $\hat{S}_z|z,\pm\rangle = \pm\frac{\hbar}{2}|z,\pm\rangle$. In this basis, the eigenvectors of the spin operator $\hat{S}_x$ are $|x,\pm\rangle = \frac{1}{\sqrt{2}}(|z,+\rangle \pm |z,-\rangle)$. Find the expectation value $\langle S_x\rangle$ for two pure states (a) $|z,+\rangle$ and (b) $|x,+\rangle$.

(a) In the basis $\{|z,+\rangle, |z,-\rangle\}$, the density operator of the pure state $|z,+\rangle$ is given by the matrix

$$\hat{\rho} = |z,+\rangle\langle z,+| \rightarrow \begin{pmatrix} 1 \\ 0 \end{pmatrix}\begin{pmatrix} 1 & 0 \end{pmatrix} = \begin{pmatrix} 1 & 0 \\ 0 & 0 \end{pmatrix} ,$$

and $\hat{S}_x$ by

$$\hat{S}_x \rightarrow \frac{\hbar}{2}\begin{pmatrix} 0 & 1 \\ 1 & 0 \end{pmatrix} .$$

Then

$$\langle S_x\rangle = \mathrm{Tr}\,\{\hat{S}_x\hat{\rho}\} = \mathrm{Tr}\left[\frac{\hbar}{2}\begin{pmatrix} 0 & 1 \\ 1 & 0 \end{pmatrix}\begin{pmatrix} 1 & 0 \\ 0 & 0 \end{pmatrix}\right] = \frac{\hbar}{2}\mathrm{Tr}\begin{pmatrix} 0 & 0 \\ 1 & 0 \end{pmatrix} = 0 \; .$$

This is exactly what we expected. Measuring $\hat{S}_x$ for a system in state $|z,+\rangle$ gives $\frac{\hbar}{2}$ and $-\frac{\hbar}{2}$ with equal probability. Note that $\rho^2 = \rho$, signaling that $|z,+\rangle$ is a pure state.

(b) In the basis $\{|z,+\rangle, |z,-\rangle\}$, the density operator of the pure state $|x,+\rangle$ is given by the matrix

$$\hat{\rho} = |x,+\rangle\langle x,+| = \frac{1}{2}(|z,+\rangle + |z,-\rangle)(\langle z,+| + \langle z,-|) \rightarrow \begin{pmatrix} 1/2 & 1/2 \\ 1/2 & 1/2 \end{pmatrix} .$$

Therefore, we obtain

$$\langle S_x \rangle = \mathrm{Tr}\,\{\hat{S}_x\hat{\rho}\} = \mathrm{Tr}\left[\frac{\hbar}{2}\begin{pmatrix} 0 & 1 \\ 1 & 0 \end{pmatrix}\begin{pmatrix} 1/2 & 1/2 \\ 1/2 & 1/2 \end{pmatrix}\right] = \frac{\hbar}{2} .$$

This result is consistent with the fact that $|x,+\rangle$ is an eigenstate of $\hat{S}_x$. Note again that $\rho^2 = \rho$. Finally, the same calculation could have been performed in $\{|x,+\rangle, |x,-\rangle\}$ basis, and we would have arrived at the same results. Proving this is left as a problem to the reader.

13.4.3.2 The density operator for mixed states

In the case of mixed states, the density operator is a weighted sum of pure state density operators, reflecting the mixture of different quantum states in the ensemble:

$$\hat{\rho} = \sum_i p_i\,\hat{\rho}_i = \sum_i p_i\,|\psi_i\rangle\langle\psi_i| \;, \tag{13.33}$$

where p_i is the probability of occurrence of the pure state $\hat{\rho}_i$. Like all probabilities, $0 \leq p_i \leq 1$, and

$$\sum_i p_i = 1 \,.$$

The information on the mixture is gathered statistically, via measurements on the pure states in the ensemble. Note that the states $|\psi_1\rangle, |\psi_2\rangle, |\psi_3\rangle, \ldots$ are not necessarily orthogonal. However, they can always be chosen to be orthogonal; here we assume that this is the case.

In the case of the mixed states, the density operator retains most of the properties from the pure states case, with the exception of the idempotence. For example, the trace of $\hat{\rho}$ is

$$\mathrm{Tr}\,\hat{\rho} = \sum_i p_i \mathrm{Tr}\,\hat{\rho}_i = \sum_i p_i = 1 \,.$$

Exercise 13.3 Prove properties 2, 3, and 4 for mixed states.

We now check under what conditions the operator $\hat{\rho}$ remains idempotent. We have

$$\begin{aligned}
\hat{\rho}^2 &= \sum_{i,j} p_i p_j \hat{\rho}_i \hat{\rho}_j \\
&= \sum_{i,j} p_i p_j \left|\psi_i\right\rangle \left\langle\psi_i|\psi_j\right\rangle \left\langle\psi_j\right| \\
&= \sum_{i} p_i^2 \left|\psi_i\right\rangle \left\langle\psi_i\right| = \sum_{i} p_i^2 \hat{\rho}_i \neq \hat{\rho} \ ,
\end{aligned}$$

where we used the fact that $|\psi_i\rangle$ are orthonormal. Note that $\hat{\rho}^2 = \hat{\rho}$ if and only if all p_i are zero with the exception of one, say p_k which is equal to one: $p_k = 1$. As result, the trace of $\hat{\rho}^2$ can be evaluated to:

$$\operatorname{Tr} \hat{\rho}^2 = \sum_i p_i^2 \operatorname{Tr} \hat{\rho}_i = \sum_i p_i^2 \leq 1 \ . \tag{13.34}$$

The trace is equal to one only for a pure state.

Example 13.2 The density matrix for unpolarized light
As an example, we will consider the density matrix for unpolarized light, which we write as a mixture of horizontally and vertically polarized states with equal probabilities.

We have only two states, which we denote by $|H\rangle$ and $|V\rangle$, with the corresponding probabilities $p_1 = p_2 = 1/2$. Then, the density operator is given by

$$\hat{\rho} = \frac{1}{2} \left|H\right\rangle \left\langle H\right| + \frac{1}{2} \left|V\right\rangle \left\langle V\right| \ ,$$

which written in basis $\{|H\rangle, |V\rangle\}$ becomes

$$\rho = \frac{1}{2} \begin{pmatrix} 1 \\ 0 \end{pmatrix} \begin{pmatrix} 1 & 0 \end{pmatrix} + \frac{1}{2} \begin{pmatrix} 0 \\ 1 \end{pmatrix} \begin{pmatrix} 0 & 1 \end{pmatrix} = \begin{pmatrix} 1/2 & 0 \\ 0 & 1/2 \end{pmatrix} .$$

We see that the density matrix is Hermitian, and

$$\operatorname{Tr} \rho = 1 \ .$$

However

$$\rho^2 = \begin{pmatrix} 1/2 & 0 \\ 0 & 1/2 \end{pmatrix}^2 = \begin{pmatrix} 1/4 & 0 \\ 0 & 1/4 \end{pmatrix} \neq \rho \ ,$$

and

$$\operatorname{Tr} \rho^2 = \frac{1}{2} < 1 \ . \tag{13.35}$$

The last equation signals that we are indeed dealing with a mixed state.

13.4.3.3 Density operator for multiparticle systems

So far, we have applied the density matrix formalism to single-particle systems. It can be equally extended to multiparticle systems. As we will see shortly, the density operator is very useful not only for describing single particle mixed states, but also individual subsystems of a composite quantum system. This latest application is quite useful in quantum computation.

To exemplify essential concepts for multiparticle systems, we start with the two-particle pure state

$$|\Psi_{AB}\rangle = |0_A\rangle\,|1_B\rangle \tag{13.36}$$

which is the first term of the singlet spin state of Eq. (13.9), which was at the heart of EPR argument on locality and realism of quantum mechanics, and a bit later, of Bell's theorem. For simplicity, in the following we reverted to the notation $|0\rangle$ representing spin-up and $|1\rangle$ for spin-down particles. As before, A denotes the electron and B the positron. The density operator for this pure two-particle state is

$$\hat{\rho}_{AB} = \big(\,|0_A\rangle\,|1_B\rangle\,\big)\big(\,\langle 0_A|\,\langle 1_B|\,\big)\ . \tag{13.37}$$

Exercise 13.4 Show that the density operator of Eq. (13.37) is idempotent and has trace 1.

Next, we choose[4] the following basis for the four-dimensional Hilbert space:

$$\{\,|0_A\rangle\,|0_B\rangle\,,|0_A\rangle\,|1_B\rangle\,,|1_A\rangle\,|0_B\rangle\,,|1_A\rangle\,|1_B\rangle\,\}\ . \tag{13.38}$$

In this basis, the density matrix for the two-particle system becomes

$$\rho_{AB} = \begin{pmatrix} 0 & 0 & 0 & 0 \\ 0 & 1 & 0 & 0 \\ 0 & 0 & 0 & 0 \\ 0 & 0 & 0 & 0 \end{pmatrix}\ . \tag{13.39}$$

Exercise 13.5 Prove Eq. (13.39)

Next, we use the density operator formalism to find the expectation value of the spin operator $\hat{S}_z^A$ of particle A (the electron). Because we are concerned only with the electron, we extend the spin operator to the two-particle state by writing it as the tensor product of $\hat{S}_z^A$ for particle A with the identity operator 1^B for particle B (the positron)

$$\hat{S}_z^{AB} \equiv \hat{S}_z^A \otimes 1^B \to \frac{\hbar}{2}\begin{pmatrix} 1 & 0 \\ 0 & -1 \end{pmatrix} \otimes \begin{pmatrix} 1 & 0 \\ 0 & 1 \end{pmatrix} = \frac{\hbar}{2}\begin{pmatrix} 1 & 0 & 0 & 0 \\ 0 & 1 & 0 & 0 \\ 0 & 0 & -1 & 0 \\ 0 & 0 & 0 & -1 \end{pmatrix},$$

[4]While the order of vectors in a basis is arbitrary, once a given order was chosen, we must keep it in all the subsequent calculations.

and then, calculate the expected value of $\hat{S}_z^A$ by taking the trace of its product with the density operator $\hat{\rho}_{AB}$

$$\langle S_z^A \rangle = \mathrm{Tr}\left\{\hat{S}_z^{AB}\,\hat{\rho}_{AB}\right\} = \mathrm{Tr}\begin{pmatrix} 0 & 0 & 0 & 0 \\ 0 & \frac{\hbar}{2} & 0 & 0 \\ 0 & 0 & 0 & 0 \\ 0 & 0 & 0 & 0 \end{pmatrix} = \frac{\hbar}{2} . \tag{13.40}$$

Since the electron (particle A) in the product state $|\Psi_{AB}\rangle$ had spin up, we expected this result. We used the explicit matrix forms of the spin and the density operators for this example to highlight an important concept: the *reduced density operator.*

In general, the reduced density operator is obtained from the full density operator ρ_{AB} by applying a *partial trace* over system B:

$$\hat{\rho}_A \equiv \mathrm{Tr}_B\{\hat{\rho}_{AB}\} . \tag{13.41}$$

The partial trace of a tensor product of operators is defined by

$$\mathrm{Tr}_B\{|a_1\rangle\langle a_2| \otimes |b_1\rangle\langle b_2|\} \equiv |a_1\rangle\langle a_2|\,\mathrm{Tr}\left\{|b_1\rangle\langle b_2|\right\} , \tag{13.42}$$

where $|a_1\rangle$ and $|a_2\rangle$ are two arbitrary states of particle A, and similarly $|b_1\rangle$ and $|b_2\rangle$ two arbitrary states of particle B. The trace $\mathrm{Tr}\left\{|b_1\rangle\langle b_2|\right\}$ is the usual trace

$$\mathrm{Tr}\left\{|b_1\rangle\langle b_2|\right\} = \langle b_2|b_1\rangle . \tag{13.43}$$

Exercise 13.6 Prove Eq. (13.43). Hint: the completeness formula may be handy here.

Using the reduced density operator, we find information about a subsystem of a multi-particle system. For example, we will recalculate the spin expectation value of $\hat{S}_z^A$ for the electron by reducing $\hat{\rho}_{AB}$. The reduced density operator is

$$\hat{\rho}_A = \mathrm{Tr}_B\{\hat{\rho}_{AB}\} = |0_A\rangle\langle 0_A|\,\mathrm{Tr}\{|1_B\rangle\langle 1_B|\} = |0_A\rangle\langle 0_A| ,$$

which, in this case is exactly the density operator for the electron considered alone. Calculating $\langle S_z^A\rangle = \mathrm{Tr}\{\hat{S}_z^A\hat{\rho}^A\}$, we obtain as expected $\langle S_z\rangle = \hbar/2$, because the electron was in a pure spin-up state.

However, the situation changes significantly when we consider more complex systems, such as the entangled state of Eq. (13.9)

$$|\Psi^-\rangle = \frac{1}{\sqrt{2}}|0_A\rangle|1_B\rangle - \frac{1}{\sqrt{2}}|1_A\rangle|0_B\rangle . \tag{13.44}$$

This is the singlet spin state of the electron (A) – positron (B) pair which was at the heart of EPR argument. The density operator corresponding to this state is

$$\begin{aligned}\hat{\rho}_{AB} &= |\Psi^-\rangle\langle\Psi^-| = \frac{1}{2}\big(|0_A\rangle|1_B\rangle - |1_A\rangle|0_B\rangle\big)\big(\langle 0_A|\langle 1_B| - \langle 1_A|\langle 0_B|\big)\\ &= \frac{1}{2}\big(|0_A\rangle|1_B\rangle\langle 0_A|\langle 1_B| - |0_A\rangle|1_B\rangle\langle 1_A|\langle 0_B| - |1_A\rangle|0_B\rangle\langle 0_A|\langle 1_B|\\ &+ |1_A\rangle|0_B\rangle\langle 1_A|\langle 0_B|\big)\,,\end{aligned} \tag{13.45}$$

with the corresponding matrix expressed in basis given by Eq. (13.38):

$$\rho_{AB} = \frac{1}{2}\begin{pmatrix} 0 & 0 & 0 & 0\\ 0 & 1 & -1 & 0\\ 0 & -1 & 1 & 0\\ 0 & 0 & 0 & 0\end{pmatrix}\,. \tag{13.46}$$

Exercise 13.7 Derive Eq. (13.46).

This is, of course, a pure state, and it can be readily checked that the matrix is idempotent: $\rho^2_{AB} = \rho_{AB}$. The off-diagonal elements of the density matrix are an indicator of the superposition of states. They are also the typical signature for entangled states.

Now, suppose we want to get information about the spin of the electron (particle A). This time, a measurement only on the electron, yields spin-up or spin-down with equal probability. Because we make measurements on one subsystem of a bigger system, we use the reduced density operator, obtained by tracing out the positron (the B particle). We obtain:

$$\begin{aligned}\hat{\rho}_A &= \mathrm{Tr}_B\{\hat{\rho}_{AB}\}\\ &= \frac{1}{2}\big(\mathrm{Tr}_B\{|0_A\rangle|1_B\rangle\langle 0_A|\langle 1_B|\} - \mathrm{Tr}_B\{|0_A\rangle|1_B\rangle\langle 1_A|\langle 0_B|\}\\ &- \mathrm{Tr}_B\{|1_A\rangle|0_B\rangle\langle 0_A|\langle 1_B|\} + \mathrm{Tr}_B\{|1_A\rangle|0_B\rangle\langle 1_A|\langle 0_B|\}\big)\\ &= \frac{1}{2}\big(|0_A\rangle\langle 0_A|\langle 1_B|1_B\rangle - |0_A\rangle\langle 1_A|\langle 0_B|1_B\rangle - |1_A\rangle\langle 0_A|\langle 1_B|0_B\rangle\\ &+ |1_A\rangle\langle 1_A|\langle 0_B|0_B\rangle\big)\\ &= \frac{1}{2}|0_A\rangle\langle 0_A| + \frac{1}{2}|1_A\rangle\langle 1_A|\,.\end{aligned} \tag{13.47}$$

This result is remarkable. The state of two entangled particles is a pure state, which means, we have maximal information on it. The pure state is known exactly. However, the first particle, when considered by itself, is in a mixed state. We no longer have maximal knowledge. In this example, the electron (particle A) behaves like a mixed state of an unpolarized beam of electrons.

13.4.3.4 The purity of a quantum state

The degree by which a quantum state is a mixed state can be described by its *purity* $\mathcal{P}$, which is defined [25] as:

$$\mathcal{P}(\hat{\rho}) = \mathrm{Tr}\left\{\hat{\rho}^2\right\} . \tag{13.48}$$

For a pure state $\hat{\rho}^2 = \hat{\rho}$, and thus $\mathcal{P} = 1$. For a mixed state, the purity will be $\mathcal{P} < 1$. In general, the purity of a quantum state satisfies

$$\frac{1}{d} \leq \mathcal{P} \leq 1 ,$$

where d is the dimension of the Hilbert space, $\mathcal{H}$, attributed to the system. The lower bound describes a completely mixed state, which, in an orthonormal basis $\{|\psi_k\rangle\}$ has the density operator

$$\hat{\rho} = \frac{1}{d}\sum_{k=1}^{d} |\psi_k\rangle \langle\psi_k| = \frac{1}{d}\mathbf{1}_d ,$$

where $\mathbf{1}_d$ is the identity operator in the d-dimensional Hilbert space. Note that for a pure state $\mathcal{P} = 1$, which indicates that a pure state spans a one-dimensional subspace of $\mathcal{H}$.

Before we conclude this section we will look at the geometrical interpretation of the purity of a quantum state, for the particular case of single qubits. For this, we return to the Bloch sphere we introduced in Sec. 13.2. A pure state was represented by a point on the Bloch sphere. By contrast, a mixed state, will correspond to a point *inside* the Bloch sphere. To see how this works, we write the corresponding density matrix ρ of a mixed quantum state in terms of the Pauli matrices. In general, a 2×2 Hermitian matrix M can be written in terms of the Pauli matrices σ_i as

$$M = n_0 I + n_1\sigma_1 + n_2\sigma_2 + n_3\sigma_3 , \tag{13.49}$$

where n_0, n_1, n_2, and n_3 are real numbers, I is the identity matrix, and

$$\sigma_1 = \sigma_x = \begin{pmatrix} 0 & 1 \\ 1 & 0 \end{pmatrix} , \quad \sigma_2 = \sigma_y = \begin{pmatrix} 0 & -i \\ i & 0 \end{pmatrix} , \quad \sigma_3 = \sigma_z = \begin{pmatrix} 1 & 0 \\ 0 & -1 \end{pmatrix} .$$

The Pauli matrices are traceless, therefore, the condition $\mathrm{Tr}\,\rho = 1$ requires $n_0 = \frac{1}{2}$. We obtain

$$\rho = \frac{1}{2}I + \vec{n}\cdot\vec{\sigma} = \frac{1}{2}(I + \vec{a}\cdot\vec{\sigma}) , \tag{13.50}$$

where the vector $\vec{a} = 2\vec{n}$ represents the quantum state, and $\vec{\sigma} \equiv (\sigma_1, \sigma_2, \sigma_3)$. Using the identity $(\vec{a}\cdot\vec{\sigma})(\vec{b}\cdot\vec{\sigma}) = (\vec{a}\cdot\vec{b})I + i(\vec{a}\times\vec{b})\cdot\vec{\sigma}$, we obtain

$$\rho^2 = \frac{1}{2}\left[\frac{1}{2}\left(1 + |\vec{a}|^2\right) I + \vec{a}\cdot\vec{\sigma}\right] . \tag{13.51}$$

Thus, the purity of the quantum state is

$$\mathcal{P} = \text{Tr}\left\{\rho^2\right\} = \frac{1}{2}\left(1 + |\vec{a}|^2\right) .$$

For a pure quantum state we have $\mathcal{P} = 1$, which requires $|\vec{a}| = 1$. That is a point on the surface of the Bloch sphere. The maximally mixed state corresponds to $\rho = \frac{1}{2}I$, or $|\vec{a}| = 0$, which is in the center of the Bloch sphere. The point corresponding to a less than maximally mixed state will be somewhere inside the Bloch sphere.

We have seen that the purity of the state is geometrically represented by the degree by which the point representing the mixed quantum state is close to the surface of the Bloch sphere. The purity of quantum states is important in practical applications such as quantum computing. The outcome of most quantum information protocols is highly sensitive to the quality of the initial state, thus the purity of the input is essential for the practical realization of quantum processors.

13.5 Elements of quantum computing

Quantum information and quantum computing are rapidly developing fields driven by recent advancements in controlling and manipulating qubits. Although quantum computing is still in its nascent stages of addressing complex problems, quantum encryption protocols are already commercially utilized due to their relatively simple implementation and demonstrated robustness. This section will start with an example of the quantum implementation of the one-time-pad encryption technique and conclude with aspects of quantum computation.

13.5.1 Quantum cryptography

13.5.1.1 One-time pad protocol

Since ancient times, sending secret messages has been of great importance for obvious reasons. Among the many cryptographic systems, one holds a particular appeal because it can be totally secure. It is the so-called one-time-pad encryption. The sender, in our case, Alice, and the recipient, Bob, share the same key, denoted by k, which is a sequence of random bits of length n. The message m that Alice wants to send must have the same length as the key. At the end of the encoding-decoding process the key is discarded, hence the name "one-time-pad protocol." Here is how it works:

- Alice encrypts her message by computing bit-by-bit the exclusive OR or XOR, defined below, between her message and the key: $x = m \oplus k$. Then, she sends the result x to Bob.

- Bob decrypts the message by computing $m = x \oplus k$.
- The key k is discarded.

Before going any further, we must clarify the language. The XOR function is a bitwise function with two inputs, a and b, and one output y. The output y is 1 when the two inputs are different. We illustrate the logical XOR function and its truth table in Fig. 13.6.

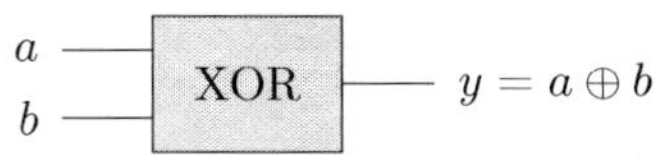

a	b	$y = a \oplus b$
0	0	0
0	1	1
1	0	1
1	1	0

FIGURE 13.6
Exclusive OR or XOR bit operation $y = a \oplus b$.

For example, we assume that Alice and Bob share the following secret key

$$k = \{0, 0, 0, 1, 1, 0, 0, 0, 0, 0, 1, 0, 0, 1, 1\} ,$$

which is 15-bit long. Thus, Alice can send a 15-bit long message to Bob. Let us assume that her message is:

$$m_A = \{0, 0, 1, 0, 0, 1, 0, 0, 0, 1, 1, 1, 1, 0, 0\} .$$

She first encrypts the message

$$x = m_A \oplus k = \{0, 0, 1, 1, 1, 1, 0, 0, 0, 1, 0, 1, 1, 1, 1\} ,$$

and sends it to Bob. Bob, uses the same key, and decrypts the message

$$m_B = x \oplus k = \{0, 0, 1, 0, 0, 1, 0, 0, 0, 1, 1, 1, 1, 0, 0\} .$$

As expected, $m_A = m_B$. If Eve, the eavesdropper, intercepts the encrypted message x, she has a meaningless sequence of random 0's and 1's. Thus, in the absence of a key, the communication is totally secure. However, the encryption could be broken if Eve obtains the secret key k. This underscores the paramount importance of keeping the encryption key secret. Because each key is discarded after each communication, the main issue to resolve is to find a secure way to distribute the key between Alice and Bob.

13.5.1.2 Quantum encryption: the BB84 protocol

The first method for secure key distribution was proposed in 1984 by Bennett and Brassard [5]. They leveraged the principles of quantum mechanics to ensure a secure communication to distribute the encryption key. This quantum key distribution protocol, known as BB84, fundamentally relies on the behavior of qubits, which, yield different results when measured in different bases. Qubits change state when observed, thus providing a clue on whether they were tampered with while in transit. This intrinsic property of qubits allows the two parties to detect any eavesdropping attempts, as any measurement on the quantum system by an unauthorized party inevitably disturbs the system and introduces detectable anomalies.

We will describe the BB84 protocol using as qubits linearly polarized photons, and two polarization bases. Following Ref. [5], we denote the rectilinear basis

$$\mathrm{R} \equiv \{|H\rangle\,, |V\rangle\}\ , \tag{13.52}$$

where the basis vectors $|H\rangle$ and $|V\rangle$ correspond to the horizontal and the vertical polarizations, respectively. Similarly, we denote the diagonal basis

$$\mathrm{D} \equiv \{|+\rangle\,, |-\rangle\}\ , \tag{13.53}$$

which is rotated by 45^0 with respect to the R basis:

$$|+\rangle = \tfrac{1}{\sqrt{2}}\left(|H\rangle + |V\rangle\right) \quad ; \quad |-\rangle = \tfrac{1}{\sqrt{2}}\left(|H\rangle - |V\rangle\right)$$

The $|+\rangle$ corresponds to the diagonal polarization and $|-\rangle$ to the anti-diagonal one, as illustrated in Fig. 13.7.

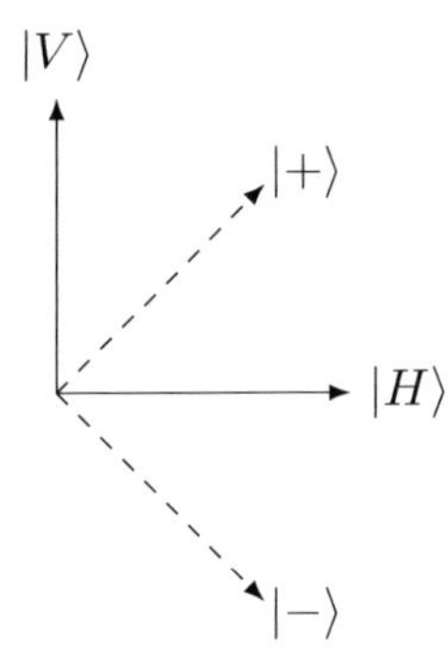

FIGURE 13.7
The rectilinear and diagonal bases for the BB84 protocol.

If Alice chooses the R basis, then she polarizes the photon horizontally for 0, and vertically for 1. Similarly, in the D basis, she polarizes the photon diagonally for 0, and anti-diagonally for 1. We are now ready to describe the full BB84 protocol.

1. Alice will randomly polarize N photons as follows:
   - She generates a random sequence of N bits which are either 0 or 1. Truly random numbers can be generated for example by measuring in a rectilinear basis the polarization of photons which were diagonally polarized. One obtains 0 or 1 with equal probability.
   - For 0, Alice intends to encode the photon in the R basis, and for 1 in the D basis. However, she will employ a random choice of encoding bases as shown below in Table 13.1.
2. Alice sends the resulting qubits to Bob via a public channel, which is not necessarily secure. This could be a fiber optic or a beam of polarized microwaves, etc.
3. Bob generates a random sequence of N bits at his end in the same manner as Alice. For 0, Bob will employ the R basis, and for 1 the D basis, in measuring the qubits received from Alice.
4. After the process, Alice and Bob talk on a classical channel, which may not be secure.
   - For each qubit received, Bob announces the basis he used (R or D), but not the result of his measurements.
   - Alice confirms when she had used the same basis as Bob.
   - For the qubits where Alice and Bob used the same basis, Bob has gotten the bit Alice intended to send.
   - The rest of the qubit measurements are discarded.
5. For truly random choices of bases, Alice and Bob will keep about half of the initial bits.
6. Assume now that Eve has access to qubits sent by Alice to Bob.
   - Eve cannot measure and resend the qubits to Bob because it's impossible for her to distinguish between the four possibilities $\{|H\rangle, |V\rangle, |+\rangle, |-\rangle\}$.
   - If Eve chooses a basis at random, she will get the correct results only about half of the time, and Alice and Bob may detect her interaction by sharing openly some of the bits in the key.
   - Eve cannot intercept and copy the qubits from Alice, because of the no-cloning theorem[5]. So she cannot have her own copy of the message and wait for the communication between Alice and Bob.

[5]The no-cloning theorem states that a quantum state cannot be duplicated. [19, 49]

TABLE 13.1
BB8 Protocol. Alice chooses a random key and a random sequence of bases to encrypt the key. Bob chooses a sequence of random bases to decrypt the message. After reconciling the transmission over classical communication channel, and checking for eavesdropping, the remaining key is secure.

QUANTUM TRANSMISSION															
Alice's random bits	1	0	0	1	1	0	0	0	0	0	1	0	0	1	1
Alice's random sending bases	R	R	D	D	D	D	R	R	R	D	R	D	D	D	D
Photons Alice sends	$\rightarrow$	$\rightarrow$	$\nearrow$	$\searrow$	$\searrow$	$\nearrow$	$\rightarrow$	$\rightarrow$	$\rightarrow$	$\nearrow$	$\uparrow$	$\nearrow$	$\nearrow$	$\searrow$	$\searrow$
Bob's random receiving bases	R	D	D	R	D	R	R	R	D	D	R	R	D	R	R
Bits as received by Bob	1	1	0	0	Fail	Fail	0	0	0	0	1	Fail	0	0	Fail
PUBLIC DISCUSSION															
Bob reports bases of received bits	R	D	D	R			R	R	D	D	R		D	R	
Alice confirms the correct bases	OK		OK		OK		OK	OK		OK	OK		OK		
Presumably shared information	1		0		1		0	0		0	1		0		
Bob reveals some key bits at random			0							0					
Alice confirms them			OK							OK					
OUTCOME															
Remaining shared secret bits	1				1		0	0			1		0		

In Table 13.1, we illustrate an example of the implementation of this protocol. There are some bits in the transmission process which are lost due to either the noise in the channel, or to the detectors' failure to detect the qubits. These are denoted as "fail" in the table. The BB84 protocol provides a robust framework for generating and distributing cryptographic keys with a high level of security, laying the groundwork for the practical implementation of quantum cryptography. More advanced quantum key distribution protocols are currently employed for encrypted communications.

13.5.2 Quantum information processing

Unlike classical computers which operate with bits that can take only two values, 0 and 1, quantum computers operate on qubits, which can take infinitely many values. In quantum information processing one manipulates qubits via unitary transformations (quantum gates), and projective transformations (measurements). These operations are modeled using quantum circuits, which are graphical representations of various operations on qubits.

13.5.2.1 Single qubits

To introduce the main concepts of quantum circuits we begin with a simple example. Starting with the qubit $|0\rangle$ we want to transform it into a superposition of basis vectors $|+\rangle = \frac{1}{\sqrt{2}}|0\rangle + \frac{1}{\sqrt{2}}|1\rangle$, and then measure the resulting qubit. This set of operations can be modeled as shown in Fig. 13.8:

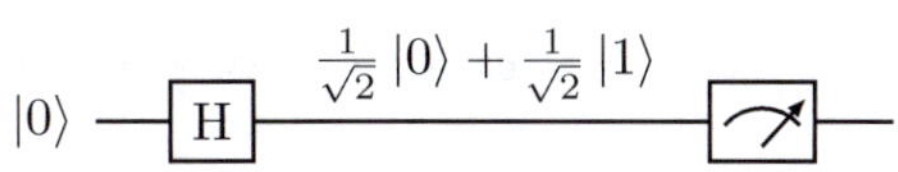

FIGURE 13.8
Using the Hadamard gate to create the balanced superposition of $|0\rangle$ and $|1\rangle$.

Let us unpack notation. We represent a qubit by a horizontal, single line, as illustrated in Fig. 13.9. The time is assumed to flow from left to right, and the convention is that, initially, all qubits are set to $|0\rangle$.

FIGURE 13.9
Quantum circuit representation of a single qubit.

The unitary transformation from $|0\rangle$ to $|+\rangle$ is modeled by the Hadamard gate, H, whose action, in the $\{|0\rangle, |1\rangle\}$ basis, is given by the matrix:

$$\mathrm{H} = \frac{1}{\sqrt{2}}\begin{pmatrix} 1 & 1 \\ 1 & -1 \end{pmatrix}. \tag{13.54}$$

We see immediately that its action on $|0\rangle$ yields

$$\mathrm{H}\,|0\rangle \rightarrow \frac{1}{\sqrt{2}}\begin{pmatrix} 1 & 1 \\ 1 & -1 \end{pmatrix}\begin{pmatrix} 1 \\ 0 \end{pmatrix} = \frac{1}{\sqrt{2}}\begin{pmatrix} 1 \\ 1 \end{pmatrix} \rightarrow \frac{1}{\sqrt{2}}\,|0\rangle + \frac{1}{\sqrt{2}}\,|1\rangle \ .$$

We notice that applying a Hadamard gate on $|0\rangle$ yields a superposition of states, which is a fundamental property used in many quantum algorithms. The last step in the circuit above is the measurement, represented by the meter symbol. In this case, one obtains with equal probability either $|0\rangle$ or $|1\rangle$. Incidentally, this simple circuit is a perfect random number generator, as it produces with uniform probability 0's and 1's. It could be used, for example, in the BB84 protocol to generate the random sequences of bases for Alice and Bob.

Exercise 13.8 Calculate the action of the Hadamard gate on an arbitrary qubit $|q\rangle = \alpha\,|0\rangle + \beta\,|1\rangle$.

In general, the "value" of any qubit is determined by measuring it. As a result of the measuring process, the qubit $|q\rangle = \alpha\,|0\rangle + \beta\,|1\rangle$ collapses at random into one of the two possible states $|0\rangle$ or $|1\rangle$, with probabilities $|\alpha|^2$ and $|\beta|^2$, respectively. In practice we will make repeated measurements on identically prepared states, thus obtaining information on the frequencies at which $|0\rangle$ and $|1\rangle$ occur.

13.5.2.2 Multiple qubits

The circuit representing two qubits consists of two horizontal lines, as shown in Fig. 13.10.

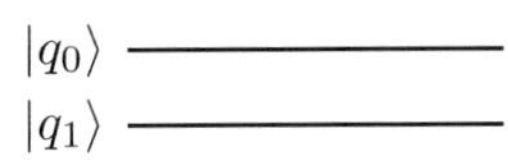

FIGURE 13.10
Quantum circuit representation of two qubits.

As mentioned earlier, the qubits are initialized on $|0\rangle$. They do not interact and are represented by the tensor product $|q_1 q_0\rangle$ as previously discussed in Sec. 13.2. Written in lexicological order, the basis for two qubits is

$$\{|00\rangle\,, |01\rangle\,, |10\rangle\,, |11\rangle\} \equiv \{|0\rangle\,, |1\rangle\,, |2\rangle\,, |3\rangle\}$$

where the RHS is the so-called *computational basis*. The labels of the vectors in the computational basis are given by their equivalent base-2 representation. For example, $|3\rangle$ corresponds to $|11\rangle$ because $3 = 1 \times 2^1 + 1 \times 2^0$.

We can readily generalize the above ideas to three or more qubits. For example, assume that we have three qubits as illustrated in Fig. 13.11.

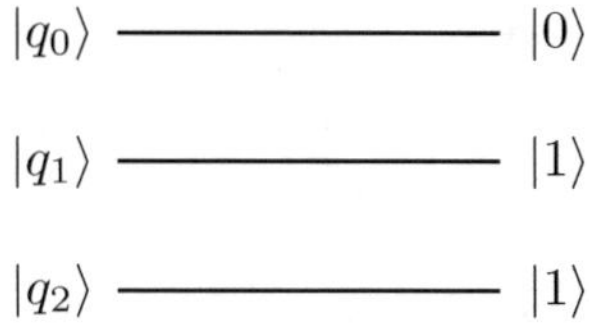

FIGURE 13.11
Quantum circuit representation the state $|6\rangle$ in the computational basis.

This is written as $|110\rangle$ or $|6\rangle$ in the computational basis. At the time of writing this book, there are publicly accessible quantum computers operating with 127 qubits.[6]

13.5.2.3 Classical bits

Quantum computing also requires the use of classical bits in addition to qubits. Classical bits can be used to collect the results of the measurements, or to control other quantum circuits. Classical bits are represented by double-lines. In the example of Fig. 13.12, we measure two qubits and collect the results on classical wires:

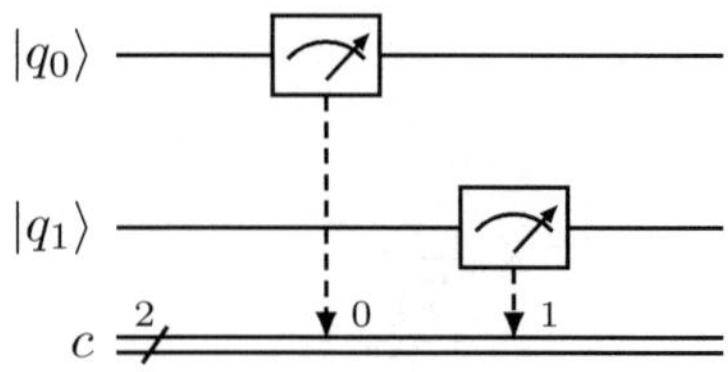

FIGURE 13.12
Measuring two qubits.

In this diagram, the result from measuring the qubit $|q_0\rangle$ is collected in the channel $c[0]$ of the classical wire, and the result of measuring the qubit $|q_1\rangle$ in $c[1]$. The diagonal line with the superscript 2 indicates the number of wires in the classical channel. In this case, we have 2 classical wires. Note that the double line representation stands regardless of the number of classical wires it contains. In the case of n classical wires, as shown in Fig. 13.13 we have

FIGURE 13.13
Quantum circuit representation of a n-bit classical channel.

[6]See for example the IBM Quantum Platform.

13.5.2.4 Single-qubit gates

Unitary transformations on qubits are modeled by quantum gates. Here are some of the commonly used single-qubit gates:

- **X-gate**. This gate flips $|0\rangle$ and $|1\rangle$. Its matrix representation is

$$\mathrm{X} = \begin{pmatrix} 0 & 1 \\ 1 & 0 \end{pmatrix},$$

  which is Pauli σ_x matrix. In quantum circuits it is represented with two equivalent symbols

FIGURE 13.14
X-gate representation.

- **Z-gate**. This gate leaves $|0\rangle$ unchanged and maps $|1\rangle$ to $-|1\rangle$. Its matrix representation is

$$\mathrm{Z} = \begin{pmatrix} 1 & 0 \\ 0 & -1 \end{pmatrix},$$

  which is Pauli σ_z matrix. In quantum circuits it is represented as

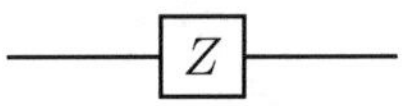

FIGURE 13.15
Z-gate representation.

- **Y-gate**. The matrix representation of the Y-gate is

$$\mathrm{Y} = \begin{pmatrix} 0 & -i \\ i & 0 \end{pmatrix},$$

  which is Pauli σ_y matrix. It is represented as

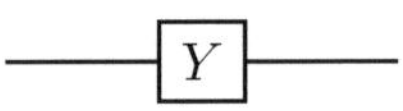

FIGURE 13.16
Y-gate representation.

$$\mathrm{H} = \frac{1}{\sqrt{2}} \begin{pmatrix} 1 & 1 \\ 1 & -1 \end{pmatrix}$$

FIGURE 13.17
H-gate matrix and quantum circuit representation.

- **Hadamard gate**. We have already encountered the Hadamard gate. Its matrix and symbols are shown in Fig. 13.17.
- **Rotation gate**. The rotation gate $R(\varphi)$ leaves $|0\rangle$ unchanged and maps $|1\rangle$ to $e^{i\varphi}|1\rangle$. Its matrix and symbol are

$$\mathrm{R}(\varphi) = \begin{pmatrix} 1 & 0 \\ 0 & e^{i\varphi} \end{pmatrix}$$

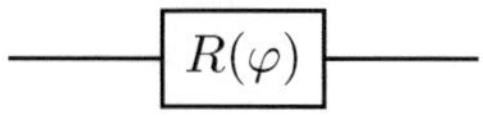

FIGURE 13.18
The quantum circuit representation of the rotation gate $R(\varphi)$.

Note that The Z-gate is $R(\pi)$.

- **The S- and T- gates**. Two gates often used are the S-gate, $S = R(\pi/2)$ and the T-gate, $T = R(\pi/4)$. Note that $Z = S^2$, so S is the square root of the Z-gate. Similarly, $S = T^2$.

13.5.2.5 Multiple-qubit gates

While we can design gates to operate on an arbitrary number of qubits, in practice we mostly encounter two- and three-qubit gates.

- **CNOT gate**. The CNOT (control-not) gate is arguably one of the most important gate in quantum computation. It operates on two qubits, called *control* and *target*. If the control qubit is $|0\rangle$, the target qubit is left unchanged. If the control qubit is $|1\rangle$, then the target qubit is flipped as if an X-gate is applied to it. The corresponding quantum circuit is illustrated below, where $|q_0\rangle$ is the control qubit and $|q_1\rangle$ is the target.

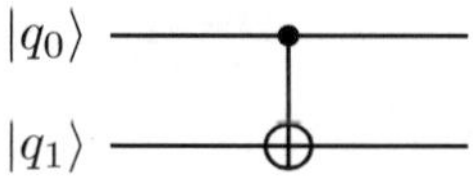

FIGURE 13.19
CNOT-gate representation.

For example, if the control qubit $|q_0\rangle$ is on $|1\rangle$, then the action of a CNOT gate on the target qubit $|q_1\rangle = \alpha|0\rangle + \beta|1\rangle$ yields

$$|q_1\rangle = \alpha|0\rangle + \beta|1\rangle \xrightarrow{\text{CNOT with control on 1}} \alpha|1\rangle + \beta|0\rangle . \tag{13.55}$$

With our convention of ordering qubits in a tensor product, the CNOT gate is represented by the following matrix

$$\mathrm{CNOT} = \begin{pmatrix} 1 & 0 & 0 & 0 \\ 0 & 0 & 0 & 1 \\ 0 & 0 & 1 & 0 \\ 0 & 1 & 0 & 0 \end{pmatrix}. \tag{13.56}$$

It is a 4×4 matrix because it operates on a tensor product of two qubits, which live in a four-dimensional Hilbert space. For example, consider the control qubit on $|1\rangle$, and the target on $|0\rangle$. Then, we expect

$$\mathrm{CNOT}\,|01\rangle = |11\rangle \tag{13.57}$$

Indeed

$$\begin{pmatrix} 1 & 0 & 0 & 0 \\ 0 & 0 & 0 & 1 \\ 0 & 0 & 1 & 0 \\ 0 & 1 & 0 & 0 \end{pmatrix} \begin{pmatrix} 0 \\ 1 \\ 0 \\ 0 \end{pmatrix} = \begin{pmatrix} 0 \\ 0 \\ 0 \\ 1 \end{pmatrix}. \tag{13.58}$$

Among other applications, we use the CNOT gate to entangle qubits, as we will show shortly.

- **SWAP.** This gates swaps the two qubits on which it is applied.

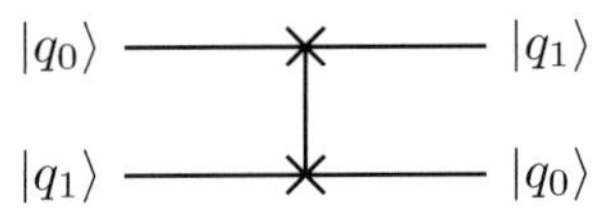

FIGURE 13.20
SWAP-gate representation.

Note that a SWAP gate can be implemented with three CNOT gates is illustrated in Fig. 13.21:

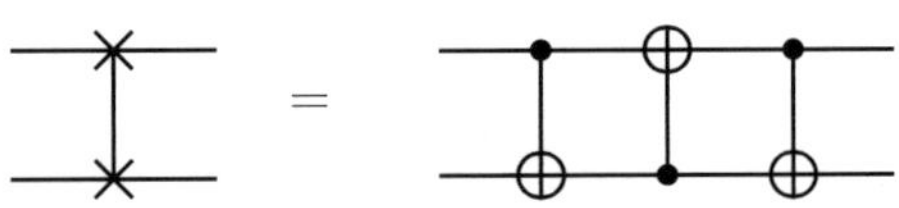

FIGURE 13.21
SWAP-gate implementation with CNOT gates.

Exercise 13.9 Find the matrix of the SWAP gate, and prove that it can be simulated with three CNOT gates as shown in the diagram above.

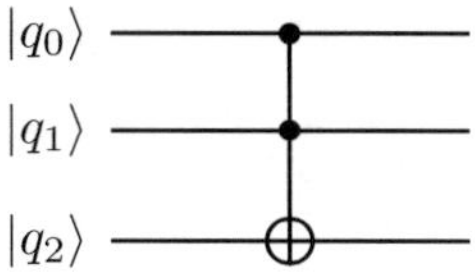

FIGURE 13.22
The Toffoli gate.

- **Toffoli**. This gate is also known as a CCNOT (control-control-not) gate. It operates on three qubits: two control qubits, $|q_0\rangle$ and $|q_1\rangle$, and one target qubit $|q_2\rangle$ as illustrated in Fig. 13.22. The third qubit will be inverted only when both of the control qubits are on $|1\rangle$.

Exercise 13.10 The Toffoli gate is a unitary operator acting on an eight-dimensional Hilbert space. Find the corresponding matrix.

13.5.3 Practical examples of quantum circuits

One freely available quantum computing platform is IBM Quantum. In the following we will explore some simple quantum circuits together with the results after running them on real quantum processors.

13.5.3.1 Superposition

We have already seen that applying a Hadamard gate on the qubit $|0\rangle$, we obtain

$$\mathrm{H}\,|0\rangle = \frac{1}{\sqrt{2}}\,|0\rangle + \frac{1}{\sqrt{2}}\,|1\rangle \ .$$

If we measure the resulting qubit we obtain either $|0\rangle$ or $|1\rangle$ with equal probability. The quantum circuit that performs this task is illustrated in Fig. 13.23. The graphical representation is slightly different from the generic one presented above because it resulted from Qiskit, an open-source language used to program IBM quantum computers.

The experiment ran on the quantum processor `ibm_sherbrooke`, for 4,096 times or "shots." Performing the measurements we obtained the result 0 for 2,075 times and 1 for 2,021 times. It is not quite equally split, but a higher number of measurements would produce a more balanced set of results. The corresponding histogram is shown in Fig. 13.24.

13.5.3.2 Entanglement

Having established a way to put two qubits in superposition, the next step is to entangle them. As an example, we will generate the following Bell state

$$|\Phi^{+}\rangle = \frac{1}{\sqrt{2}}\,|00\rangle + \frac{1}{\sqrt{2}}\,|11\rangle \ . \tag{13.59}$$

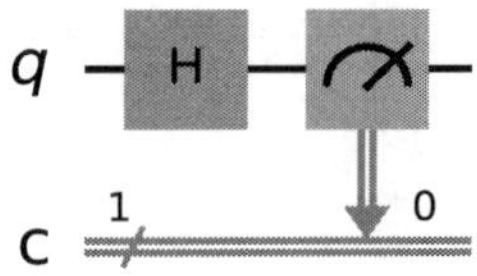

FIGURE 13.23
Superposition.

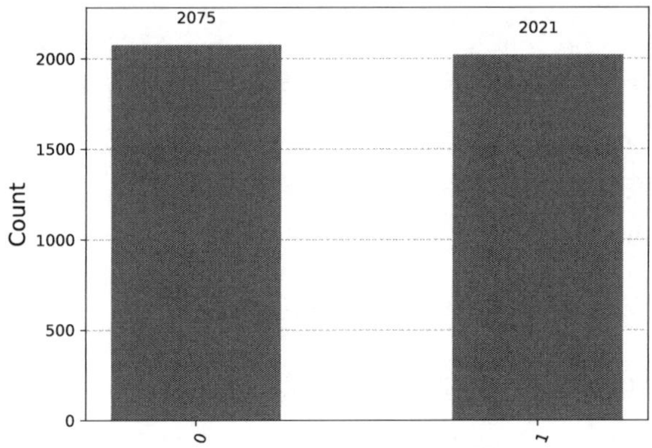

FIGURE 13.24
The histogram of 4,096 measurements of the quantum state $|+\rangle = \frac{1}{\sqrt{2}}|0\rangle + \frac{1}{\sqrt{2}}|1\rangle$. Note the almost equal distribution of 0 and 1.

We need a quantum circuit with two quantum channels (one for each qubit) and two classical channels, for measurements. The circuit is illustrated in Fig. 13.25. Initially, both qubits are set at $|0\rangle$. The state of the system is thus

$$|\Psi_0\rangle = |q_1 q_0\rangle = |00\rangle \ .$$

Remember that we read the qubits in a circuit from bottom-up, so qubit $|q_1\rangle$ is the first term in the tensor product. In the next step, we apply a Hadamard gate on the qubit $|q_0\rangle$ which becomes $|+\rangle$. Consequently, the state of the system becomes

$$|\Psi_1\rangle = |0\rangle |+\rangle = |0\rangle \left(\frac{1}{\sqrt{2}} |0\rangle + \frac{1}{\sqrt{2}} |1\rangle \right)$$

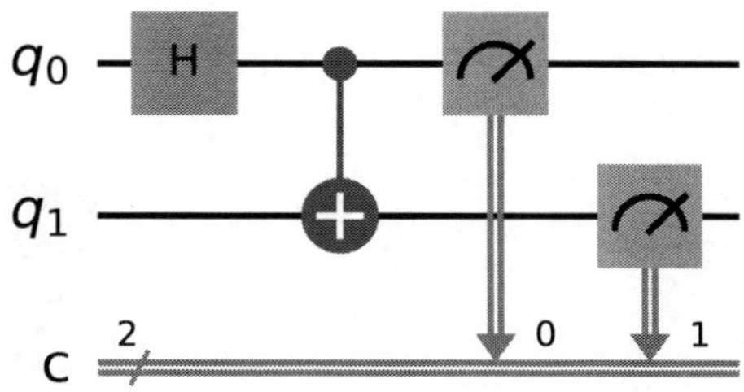

FIGURE 13.25
Quantum circuit that generates the Bell $|\Phi^+\rangle$ entangled state.

At this step, the qubit $|q_0\rangle = |+\rangle$ is used as control qubit for a CNOT gate with target $|q_1\rangle = |0\rangle$. CNOT is a linear unitary gate, therefore, when the control qubit is $|+\rangle$, its action on the target will be a superposition of the actions of control $|0\rangle$ –which does nothing to the target– with control $|1\rangle$ –which flips the target. Thus, the state of the system becomes

$$|\Psi_2\rangle = \frac{1}{\sqrt{2}} |0\rangle |0\rangle + \frac{1}{\sqrt{2}} |1\rangle |1\rangle$$

which is exactly the state $|\Phi^+\rangle$ of Eq. (13.59). Subsequent measurements on this state should yield either 00 or 11 with equal probability, but no 01 or 10. Running this circuit for 4,096 shots on the quantum processor `ibm_osaka` [7] we obtained:

TABLE 13.2
The counts for measuring the entangled qubits of the Bell state $|\Phi^+\rangle$ as implemented on a real quantum computer.

Measurement	00	01	10	11
Counts	1995	66	141	1894

The corresponding histogram is shown in Fig. 13.26.

Note that the "forbidden" readings of 01 and 10 appear, but with very low counts. These are errors which are likely caused by the thermal noise in the quantum processor unit.

13.5.3.3 Teleportation

Quantum teleportation is, unfortunately, not quite as spectacular as shown in many science-fiction movies. We do not have a person disappearing from one location only to appear in another distant location using some mysteriously buzzing technology.

[7] IBM is rapidly modifying its quantum processors. Shortly after running this experiment, `ibm_osaka` was decommissioned.

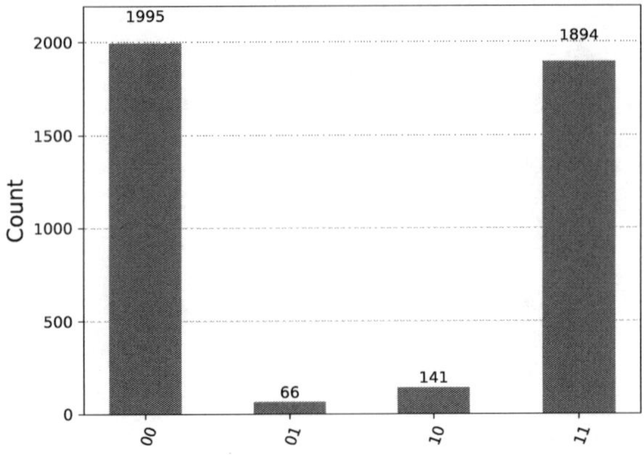

FIGURE 13.26
The histogram of 4,096 measurements of the entangled quantum state $|\Phi^+\rangle$.

In the current practice, teleportation is understood as a procedure for sending a particle's quantum state from A to B without sending the particle itself. Alice, the physicist located at position A, wants to send to Bob, located at B, all the information necessary so that Bob can reconstruct at B the qubit $|q\rangle = \alpha\,|0\rangle + \beta\,|1\rangle$ initially situated at A. In the process, the qubit disappears from Alice's location and reappears at Bob's. In a way, it resembles the science-fiction depiction of teleportation.

A simple method to send a qubit $|q\rangle = \alpha\,|0\rangle + \beta\,|1\rangle$ from A to B is to send the qubit itself. However, we assume that this is not possible. Alice cannot access the information about the amplitudes α and β because if she measured the qubit, it would irreversibly collapse it into either $|0\rangle$ or $|1\rangle$. Also, the no-cloning theorem forbids her from making many copies of $|q\rangle$ in order to measure the resulting identical systems, and thus find some information about the qubit by determining the probabilities $|\alpha|^2$ and $|\beta|^2$.

This conundrum can be solved using two tenets of quantum states: superposition and entanglement. Suppose that Alice and Bob each have one particle of an entangled state, or an EPR pair. This can be achieved in several ways. For example, they entangle a pair of particles and each of them retain one of the particles. Then, Bob takes his particle away, and will use it to complete the teleportation when he arrives at B. Similarly, Alice and Bob can subscribe to a EPR service system that sends out pairs of entangled particles to their respective locations.

We will not concern ourselves with the concrete way in which Alice and Bob get their EPR particles. We will further assume that their EPR pair is the Bell state:

$$|\Phi^+\rangle_{BA} = \frac{1}{\sqrt{2}}|0\rangle_B|0\rangle_A + \frac{1}{\sqrt{2}}|1\rangle_B|1\rangle_A\ , \tag{13.60}$$

where we added the indices to specify the location of each qubit. We denote the qubit that Alice wants teleported to Bob by

$$|\psi\rangle_A = \alpha|0\rangle_A + \beta|1\rangle_A\ . \tag{13.61}$$

When the teleportation process is complete, Bob will end up with the quantum state

$$|\psi\rangle_B = \alpha|0\rangle_B + \beta|1\rangle_B\ , \tag{13.62}$$

which is the qubit Alice wanted to teleport.

One way of implementing the teleportation protocol is the following:

- *Step 1: Preparation.* In this step, Alice manipulates the qubits to which she has access, which we labeled with index A, to bring the system from the initial state $|\psi_0\rangle$ to an intermediate state $|\psi'\rangle$. Here

$$|\psi_0\rangle \equiv |\Phi^+\rangle_{BA}|\psi\rangle_A = \frac{1}{\sqrt{2}}(|0\rangle_B|0\rangle_A + |1\rangle_B|1\rangle_A)(\alpha|0\rangle_A + \beta|1\rangle_A)\ ,$$

and

$$|\psi'\rangle \equiv \frac{|T_1\rangle_B|0\rangle_A|0\rangle_A + |T_2\rangle_B|0\rangle_A|1\rangle_A + |T_3\rangle_B|1\rangle_A|0\rangle_A + |T_4\rangle_B|0\rangle_A|0\rangle_A}{2}\ ,$$

where

$$\begin{aligned}|T_1\rangle_B &\equiv \alpha|0\rangle_B + \beta|1\rangle_B\\ |T_2\rangle_B &\equiv \alpha|0\rangle_B - \beta|1\rangle_B\\ |T_3\rangle_B &\equiv \alpha|1\rangle_B + \beta|0\rangle_B\\ |T_4\rangle_B &\equiv \alpha|1\rangle_B - \beta|0\rangle_B\ .\end{aligned}$$

We will soon explain how the state $|\Psi'\rangle$ is created. Note that in this intermediate state, due to entanglement, the qubit at Bob's location acquired the unknown amplitudes α and β, as part of four possible teleported states: $|T_1\rangle\ , |T_2\rangle\ , |T_3\rangle$, or $|T_4\rangle$.

- *Step 2: Measurement.* At this stage Alice measures her two qubits, and obtains one of the four possible results: 00, 01, 10, or 11. As a result, the system's state collapses accordingly: if she obtained 00, then the system collapsed into $|T_1\rangle_B|0\rangle_A|0\rangle_A$; for 01, it collapsed into $|T_2\rangle_B|0\rangle_A|1\rangle_A$, and so on. Consequently, after Alice's measurement, due to entanglement, the qubit at Bob's location collapsed into either $|T_1\rangle$ if Alice measured 00, or in $|T_2\rangle$ if she measured 01, etc. However, Bob has no way of knowing which of the four possible teleportation states he obtained. He must "talk" to Alice to find out the results of her measurements.

- *Step 3: Communication.* Alice uses a classical channel to inform Bob of the results of her measurements. This can be done via radio, or she can write a letter and mail it to him.
- *Step 4: Final state reconstruction.* Having received the communication from Alice, Bob can recover the original quantum state using his qubit. Specifically, if the message from Alice was:

  00: Bob knows that his qubit collapsed into $|T_1\rangle_B = \alpha|0\rangle_B + \beta|1\rangle_B$, which is precisely the state that Alice intended to teleport. In this case there is nothing left to do. Bob has the final state.

  01: Bob knows that his qubit collapsed into $|T_2\rangle_B = \alpha|0\rangle_B - \beta|1\rangle_B$. In this case he applies the Z gate on his qubit and recovers the original state. Remember that the Z gate leaves $|0\rangle$ unchanged, and changes $|1\rangle$ to $-|1\rangle$.

  10: Bob knows that this case corresponds to the third possibility $|T_3\rangle_B = \alpha|1\rangle_B + \beta|0\rangle_B$, and he applies an X gate to his particle.

  11: This corresponds to $|T_4\rangle_B = \alpha|1\rangle_B - \beta|0\rangle_B$, and Bob applies an X gate followed by a Z gate to his qubit.

  Note that at the end of the process, Alice's qubit $\alpha|0\rangle_A + \beta|1\rangle_A$ was reconstructed at Bob's location $\alpha|0\rangle_B + \beta|1\rangle_B$, without Alice or Bob having any knowledge of the amplitudes α and β.

The quantum circuit that perform these operations is illustrated in Fig. 13.27. Alice has access to qubits q_0 and q_1, and Bob to qubit q_2. The qubit to

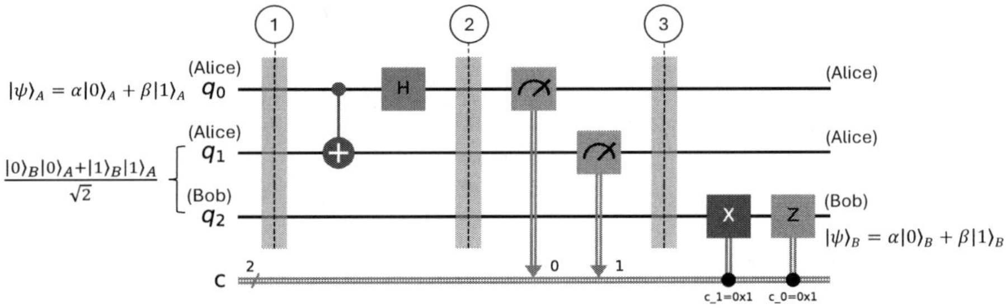

FIGURE 13.27
Teleportation circuit.

be teleported is initially on position q_0 in the circuit. Qubits q_1 and q_2 are already entangled in Bell's state $|\Phi^+\rangle$. To guide the description of the circuit we introduced three-time markers labeled ①, ②, and ③.

1. In first stage, marked by ①, the state of the system is

$$|\psi_0\rangle = |\Phi^+\rangle_{BA}|\psi\rangle_A = \frac{|0\rangle_B|0\rangle_A + |1\rangle_B|1\rangle_A}{\sqrt{2}}\left(\alpha|0\rangle_A + \beta|1\rangle_A\right)$$

2. In stage ② Alice used the qubit she wants to teleport as the control qubit of a CNOT gate with the target on her particle from the EPR pair. Then she applied a Hadamard gate on q_0. Consequently, the state of the system became

$$\begin{aligned} |\psi'\rangle &= \frac{\alpha}{\sqrt{2}}(|0\rangle_B |0\rangle_A + |1\rangle_B |1\rangle_A)\frac{|0\rangle_A + |1\rangle_A}{\sqrt{2}} \\ &+ \frac{\beta}{\sqrt{2}}(|0\rangle_B |1\rangle_A + |1\rangle_B |0\rangle_A)\frac{|0\rangle_A - |1\rangle_A}{\sqrt{2}} . \end{aligned}$$

   Rearranging the terms, we obtain

$$\begin{aligned} |\psi'\rangle &= \frac{1}{2}\Big[(\alpha|0\rangle_B + \beta|1\rangle_B)|0\rangle_A|0\rangle_A + (\alpha|0\rangle_B - \beta|1\rangle_B)|0\rangle_A|1\rangle_A \\ &+ (\alpha|1\rangle_B + \beta|0\rangle_B)|1\rangle_A|0\rangle_A + (\alpha|1\rangle_B - \beta|0\rangle_B)|1\rangle_A|1\rangle_A\Big] \\ &= \frac{|T_1\rangle_B|0\rangle_A|0\rangle_A + |T_2\rangle_B|0\rangle_A|1\rangle_A + |T_3\rangle_B|1\rangle_A|0\rangle_A + |T_4\rangle_B|0\rangle_A|0\rangle_A}{2} \end{aligned}$$

   which is exactly the intermediate state we were looking for.

3. In between ② and ③ Alice measures q_0 and q_1 and sends the results on the classical channel to Bob. The result of measuring q_0 is sent through the first bit of the channel i.e., $c[0]$, and the result from q_1 through the second bit, $c[1]$.

4. Bob employs the classically-controlled gate X followed by the classically-controlled gate Z, to recover the initial qubit. These gates are activated only when the corresponding control bit is on 1.

Exercise 13.11 Calculate explicitly the system states in steps 1-4 above.

Thus, the initial state $|\psi\rangle_A$ was teleported from Alice to Bob. Of course, depending on the particular implementation of the circuit, the initial EPR pair could be another Bell state, such as $|\Phi^-\rangle$ or $|\Psi^\pm\rangle$.

13.6 Problems

13.1 In the Schrödinger picture state vectors are time dependent while the operators are not. In the Heisenberg picture, the image is reversed. Operators are time dependent while the state vectors become time independent. Show that the density operator is time independent in the Heisenberg picture.

13.2 Find the density matrix for unpolarized light using an equal mixture of left and right-circularly polarized photons, and show that we arrive at the same result as in Eq. (13.35).

13.3 Consider the density operator $\hat{\rho} = \frac{1}{2}|z,+\rangle\langle z,+| + \frac{1}{2}|y,+\rangle\langle y,+|$ where $|y,\pm\rangle = \frac{1}{\sqrt{2}}(|z,+\rangle \pm i\,|z,-\rangle)$ are the eigenvectors of the $\hat{S}_y$ spin component. Construct the density operator in the $\{|z,+\rangle, |z,-\rangle\}$ basis and show that it describes a mixed state. Calculate $\langle S_y\rangle$ and $\langle S_x\rangle$ in this state.

13.4 For each of the four Bell states of Eq. (13.8) find the reduced density operator for the first qubit. Repeat the calculations for the second qubit only for $|\Phi^+\rangle$ and $|\Psi^+\rangle$.

13.5 Find the density operator for the pure state given by Eq. (13.7) if $\alpha, \beta, \gamma, \delta \in \mathbb{R}$. Calculate the reduced densities for the constituent single qubits by taking the corresponding partial traces.

13.6 Prove Eq. (13.51).

13.7 Show that:

$$
\begin{aligned}
&HXH = Z \\
&HZH = X \\
&HYH = -Y \\
&Y = iXZ \\
&H^\dagger = H = H^{-1} \\
&H = (X+Z)/\sqrt{2} \\
&X^2 = Y^2 = Z^2 = H^2 = 1_{2\times 2}
\end{aligned}
$$

13.8 Build a quantum circuit for a teleportation protocol that uses the Bell state $|\Psi^-\rangle$ as the initial EPR pair.

Bibliography

[1] M. Abramowitz and I. A. Stegun, editors. Handbook of Mathematical Functions. Dover, New York, 9th edition, 1965.

[2] A. Aspect, P. Grangier, and G. Roger. Experimental Realization of Einstein-Podolsky-Rosen-Bohm Gedankenexperiment: A New Violation of Bell's Inequalities. Phys. Rev. Lett., 49:91–94, Jul 1982.

[3] M. Beck. Quantum Mechanics: Theory and Experiment. Oxford University Press, 2012.

[4] J. S. Bell. On the Einstein Podolsky Rosen paradox. Physics, 1(3):195–200, 1964.

[5] C. H. Bennett and G. Brassard. Quantum Cryptography: Public Key Distribution and Coin Tossing. In International Conference of Computers, Systems & Signal Processing, volume 1, pages 175–179, Bangalore, India, December 1984. IEEE.

[6] A. Berera and L. Del Debio. Quantum Mechanics. Cambridge University Press, 2022.

[7] D. Bohm. A Suggested Interpretation of the Quantum Theory in Terms of "Hidden" Variables. I. Physical Review, 85:166–179, 1952.

[8] D. Bohm. A Suggested Interpretation of the Quantum Theory in Terms of "Hidden" Variables. II. Physical Review, 85:180–193, 1952.

[9] D. Bohm and Y. Aharonov. Discussion of Experimental Proof for the Paradox of Einstein, Rosen, and Podolsky. Phys. Rev., 108:1070–1076, Nov 1957.

[10] C. Cohen-Tannoudji, B. Diu, and F. Laloë. Quantum Mechanics, Volume 1: Basic Concepts, Tools, and Applications. Wiley, 2019.

[11] P. A. M. Dirac. The Principles of Quantum Mechanics. Oxford University Press, Oxford, 4th edition, 1958.

[12] A. Einstein, B. Podolsky, and N. Rosen. Can quantum-mechanical description of physical reality be considered complete? Physical Review, (47):777–780, 1935.

[13] U. Fano. Description of States in Quantum Mechanics by Density Matrix and Operator Techniques. Rev. Mod. Phys., 29:74–93, Jan 1957.

[14] R. P. Feynman, R. B. Leighton, and M. L. Sands. Electromagnetism and Matter. Basic Books, 2nd edition, 2010.

[15] A. Gangopadhyaya and G. Ramsey. Unintended Consequences of Imprecise Notation – an Example from Mechanics. American Journal of Physics, 81(4):313–315, 2013.

[16] H. Goldstein. Classical Mechanics. Pearson Education, 2002.

[17] I. S. Grant and W. R. Phillips. Electromagnetism. Wiley, 2nd edition, 1990.

[18] D. J. Griffiths. Introduction to Electrodynamics. Cambridge University, 5th edition, 2024.

[19] D. J. Griffiths and D. F. Schroeter. Introduction to Quantum Mechanics. Cambridge University Press, Cambridge, 3rd edition, 2018.

[20] P. Hamill. A Student's Guide to Lagrangians and Hamiltonians. Cambridge University Press, 2014.

[21] L. Hand and J. Finch. Analytical Mechanics. Cambridge University Press, 1998.

[22] M. A. Heald and J. B. Marion. Classical Electromagnetic Radiation. Cambridge University Press, Cambridge, 3rd edition, 2012.

[23] T. Imbo and U. Sukhatme. Logarithmic perturbation expansions in nonrelativistic quantum mechanics. American Journal of Physics, 53(2):140–146, 1984.

[24] F. J. Milford J. R. Reitz and R. W. Christy. Foundations of Electromagnetic Theory. Pearson, 4th edition, 2009.

[25] G. Jaeger. Quantum Information—An Overview. Springer, 2010.

[26] T. W. B. Kibble and F. H. Berkshire. Classical Mechanics. Imperial College Press, 2004.

[27] D. Kleppner and R. Kolenkow. An Introduction to Mechanics. McGraw-Hill, 1973.

[28] L. D. Landau and E. M. Lifchitz. Quantum Mechanics: Non-Relativistic Theory. Pergamon, Oxford, 3rd edition, 1977.

[29] N. A. Lemos. Comment on "Unintended consequences of imprecise notation: An example from mechanics", [Am. J. Phys. 81, 313–315 (2013)]. American Journal of Physics, 82:164–165, 2014.

[30] R. L. Liboff. Introductory Quantum Mechanics. Addison-Wesley, 2003.

[31] J. Marion and S. Thornton. Classical Dynamics of Particles and Systems. Brooks/Cole, 5th edition, 1986.

[32] D. Morin. Introduction to Classical Mechanics: With Problems and Solutions. Cambridge University Press, 2005.

[33] H. J. W. Müller-Kirsten. Electrodynamics. World Scientific, 2nd edition, 2011.

[34] H. J. W. Müller-Kirsten. Classical Mechanics and Relativity. World Scientific, 2nd edition, 2024.

[35] E. F. Nichols and G. F. Hull. The Pressure due to Radiation. The Astrophysical Journal, 17(5):315–351, 1903.

[36] M. A. Nielsen and I. L. Chuang. Quantum Computation and Quantum Information: 10th Anniversary Edition. Cambridge University Press, 2011.

[37] H. Ohanian. Classical Electrodynamics. Jones and Bartlett, 2nd edition, 2006.

[38] W. Pauli. Electrodynamics. Dover, New York, 2000.

[39] G. L. Pollack and D. R. Stump. Electromagnetism. Addison-Wesley, 2002.

[40] E. M. Purcell and D. J. Morin. Electricity and Magnetism. Cambridge University, 3rd edition, 2013.

[41] C. Rasinariu and A. Gangopadhyaya. The motion of two identical masses connected by an ideal string symmetrically placed over a corner. American Journal of Physics, 84(1):836–840, 2016.

[42] R. W. Robinett. Quantum Mechanics: classical results, modern systems, and visualized examples. Oxford University Press, Oxford, 2006.

[43] J. J. Sakurai and J. J. Napolitano. Modern Quantum Mechanics. Cambridge University Press, Cambridge, 2nd edition, 2017.

[44] W. M. Saslow. Electricity Magnetism and Light. Brooks/Cole, 2002.

[45] L. I. Schiff. Quantum Mechanics. McGraw-Hill, 1986.

[46] M. Schwartz. Principles of Electrodynamics. Dover, New York, 1987.

[47] I. E. Tamm. Fundamentals of the Theory of Electricity. Mir Publishers, 9th edition, 1979.

[48] J. Taylor. Classical Mechanics. University Science Books, 2005.

[49] J. S. Townsend. A Modern Approach to Quantum Mechanics. University Science Books, 2nd edition, 2012.

[50] R. K. Wangsness. Electromagnetic Fields. John Wiley and Sons, 2nd edition, 1986.

[51] S. Weinberg. Lectures on Quantum Mechanics. Cambridge University Press, 2nd edition, 2015.

Index